Science at Home
(Nature and Science Readers)

**Edith M. Patch , Harrison E. Howe and
George M. Richards (Illustrator)**

CONTENTS

Part One. Flower Gardens

PAGE

WILD FLOWERS IN GARDENS 3

The Purple Foxglove 5
Snapdragons 8
Butter and Eggs 9
Mullein 12
Turtlehead 16
Questions and Activities 17

GARDENS FOR GUESTS 20

Lilacs for Butterflies 21
Lilacs for Clearwings 23
Petunias for Humming-Bird Moths 25
Pollen and Nectar for Bees . . . 29
Leaf-Cutter Bee 30
Red Flowers for Humming Birds . 32
Berries for Birds 35
Questions and Activities 36

PART TWO. HOME SHELTERS

PAGE

DIFFERENT KINDS OF WALLS 41

Shelters of Hides 41

Shelters of Snow and Ice 42

Shelters of Plant Materials . . . 43

Shelters of Earth Materials . . . 50

Rammed-Earth Walls 55

Questions and Activities 58

THE MASON'S WORK 61

Bird and Insect Masons 61

Buildings of Stone 65

Buildings of Brick 70

Mortar and Concrete 73

The Mason's Tools 76

Questions and Activities 79

WOOD AND THE CARPENTER 81

Bird and Insect Carpenters . . . 81

Log Cabins 84

PAGE

Preparing and Using Lumber . . 86

Building Boards 91

The Mason Helps the Carpenter . 93

Wooden Ornaments 94

Saving Some Trees 95

Questions and Activities 98

WATCHING THE PAINTER 102

Protecting Metal Parts 102

Unpainted Wood 104

Paints for Wood 107

Pigments 110

Linseed Oil 111

Driers and Thinners 113

The Painter Indoors 115

Varnish 116

Shellac and Lacquer 119

Wax 121

Natural Uses of Painters' Materials 121

Questions and Activities 127

PART THREE. WATER, LIGHT, HEAT

PAGE

THE PLUMBER 133

Water for Country Houses . . . 135

Wastes from Country Houses . . 147

City Waterworks 149

City Sewage 153

Questions and Activities 156

THE ELECTRICIAN COMES 157

Electric Lights and Flame Lights . 157

Electric Shocks 161

Insulation 163

Switches and Fuses 164

Placing the Wires 167

The Electric Servant 170

Electricity in Animals 171

Questions and Activities 175

AIR: HOT AND COLD 178

Normal Body Temperature . . . 178

PAGE

Hot Air for Winter 181

How a Thermometer Works . . . 183

Wood and Charcoal 187

Coal and Coke 190

Gas 194

Oil 195

Electricity 195

Stoves and Furnaces 197

Ventilation 203

Comfortable Air in Houses . . . 204

Cooling by Evaporation 204

Questions and Activities 205

PART FOUR. THE HOME AS A WORKSHOP

CARPENTER TOOLS 211

Claw Hammer for Pounding and
  Pulling 212

Snips or Shears 218

PAGE

Wood-cutting Tools and Chips . . 220

Some Screws 224

Questions and Activities 227

A COOK'S TOOLS 229

The Name *Kitchen* 229

Measuring Devices 231

Three Classes of Levers 234

Screws Again 235

Cranks 236

Dishes to Stand Heat and Cold . 238

Discolored Metals 240

Questions and Activities 243

MATERIALS AND DEVICES FOR CLEANING 247

Soap 248

Scouring Materials 250

Brooms and Brushes 253

Suction, or Vacuum, Cleaners . . 258

Questions and Activities 262

PART FIVE. MATERIALS FOR CLOTHES

PAGE

COTTON 267

Cotton Trees and Shrubs 267

The Cotton Belt 269

Several Kinds of Cotton 270

Cotton Fibers 273

Cotton Gin 276

The Cotton Trade 277

Diseases of Cotton Plants 278

Insects That Trouble Growers of

Cotton 281

Helpful Insects 287

Questions and Activities 289

LINEN 291

Pure Irish Linen 291

Common Flax 293

Fiber Flax and Seed Flax 295

Bacteria Help in the Retting Process 297

PAGE

Flax Wilt 298

Flax Seeds from South America . . 299

Questions and Activities 302

SILK 304

Spider Silk 304

Silk Glands and Spinnerets of
Spiders 306

Caterpillar Silk 308

Wild Silk 310

The Mulberry Silkworm 312

Silk Glands and Spinnerets of Caterpillars 318

Questions and Activities 319

RAYON 321

Cellulose 323

Men Make Fibers from Liquid Cellulose 325

Four Processes 326

PAGE

Hardening the Fibers 328

Cotton Linters and Wood Pulp . . 329

Rayon from Spruce Logs 331

The Name *Rayon* 335

Questions and Activities 337

FUR AND WOOL 338

Clothes for Warmth 338

Fur for Show 341

Trapping Fur Bearers 342

Humane Fur Laws 345

Kindness to Animals 346

Grey Owl 347

Fur Farms 349

Wool 353

Questions and Activities 359

PART SIX. FOOD

FUEL FOODS 365

Fuel for Heat and Other Energy . 366

Sugars and Starches 368

PAGE

Fats 378

Proteins Are Fuel Foods 381

Questions and Activities 383

BODY BUILDERS 385

Proteins Are Body Builders . . . 386

Some Minerals Are Body Builders . 391

Two Minerals in Bones 392

Iron 395

Iodine 396

Water 397

Questions and Activities 399

VITAMINS 401

Vitamin A 404

Vitamin B 408

Vitamin C 410

Vitamin D 418

Questions and Activities 421

BOOKS TO READ 424

INDEX 427

A FEW WORDS ABOUT THIS BOOK

Do you enjoy a visit to a museum? Do you like to learn about the things that are put into glass cases for you to see?

We do not place our most common things in a museum. We do not make exhibits of them. We use them. The things we use every day, however, are quite as interesting as exhibits in museums. You may have noticed how pleasant it is to learn about familiar things. They seem more remarkable if you know facts about them.

Perhaps you have been curious about some of the objects in your own home. You may have wondered what they are made of or where they came from or how they were put together. It is quite likely, indeed, that you have asked more questions about such matters than people have had time to answer. It really takes a great deal of time to answer certain questions.

You may like to see if you find answers

to some of your questions in this book. Of course all the answers will not be on these pages. Any wide-awake boy or girl can ask about more subjects than can be crowded into one book.

Perhaps, however, you can answer some of your own questions. Do you think it would be rather good fun to find out some facts for yourself? There are suggestions in this book that will help you hunt for facts. You will find them on the " Questions and Activities " pages. You may think of other ways to hunt for facts, too.

Homes and objects in homes are in common everyday use. This book tells about such things. So it is called *Science at Home.* You may find that your own home is somewhat like a museum — full of interesting things.

EDITH M. PATCH

HARRISON E. HOWE

FLOWER · GARDENS

PART ONE

Foxgloves

WILD FLOWERS IN GARDENS

A garden may be a very important part of a home. Perhaps you have met some one who likes his garden even better than his house.

Indeed, a garden may seem like an outdoor living room if it is bordered by a hedge or a fence or a wall. Such a room is likely to have a carpet of soft grass. It may be furnished with a table, a few chairs, a hammock, and perhaps a swing. It needs no pictures in frames because the garden itself is a picture. Instead of having cut flowers in vases, it is decorated with unpicked blossoms. There is no plastered ceiling — there is the blue sky overhead.

Are you one of the people who like flower gardens? Have you yourself taken care of plants? Is your favorite sort of garden one in which wild flowers have been placed?

It is an interesting game to move a wild violet or other wild plant into your garden. Of course it is not fair to the plant to do this unless you can keep it in a flourishing condition.

So you will wish to notice the place where it was thriving. Was the ground wet or well drained? Was the spot shady or sunny? Was the soil clay or sand or leaf mold? You will study such matters as these if you wish to give your plant a garden home where it can be as healthy as it was before you adopted it.

Children often find great delight in the game of playing with plants in gardens. You may have noticed that grown people enjoy this same game. And you may be interested to learn about some of the plants that men have moved from wild homes into gardens.

Some of these plants have not had long journeys. They may have been carried from a neighboring meadow or roadside. Others have been fetched from far places of the earth — as from regions along rivers in Africa, from mountains in Asia, or from fields in Europe.

The Purple Foxglove

Do you like to learn a few Latin words? *Digitus* is the Latin word that means a finger or a toe. So we sometimes call our fingers and toes *digits*.

You could slip a finger into a foxglove blossom, and it is rather interesting to know that one name for this plant is *Digitalis*.

Many foxglove blossoms grow on a single spike as hollyhocks and larkspurs do. These plants are wild in western and central Europe. They are abundant on rocky hillsides. If you should visit the British Isles some year, we hope you can go to the mountains of north

A home of wild foxgloves on mountain sides in north Wales

Wales in June. At that time of year the high slopes there are purple as far as one can see. That is because so many foxgloves are in blossom.

Of course you need not go so far as north Wales to see purple foxgloves. They grow in many American flower gardens. Perhaps there are some in a garden near your own home. We call them "cultivated" instead of "wild" when they are planted in gardens. But they are the same kind of plant whether they grow on rocky hillsides or in garden borders. Men found them growing wild and liked them so well that they began to put them wherever they wished to see them. They brought some of them across the Atlantic Ocean to America.

The foxglove belongs to the Figwort Family. It is quite likely that you know some other members of this family of plants. We shall mention snapdragons, butter

Courtesy "Nature Magazine"

Snapdragons

and eggs, mullein, and turtleheads — some well-known relatives of the fox-gloves.

Snapdragons

Snapdragons have also been brought to gardens from certain places in Europe where they grow wild. They are frequently grown in green-houses, too, so that their flowers may be enjoyed during the winter.

Children usually like to play with snap-dragons. You may have learned how to make one of these blossoms open its mouth by pressing lightly against its sides with your thumb and finger. Then when you take away your hand, the "dragon's" mouth shuts with a quick snap.

A bumblebee can open a snapdragon when she desires to reach the nectar that the blossom holds for her. She has a very clever way of doing this.

Butter and Eggs

The blossoms of the plant called butter and eggs are shaped very much like those of their relative the snapdragon. Their mouths can be made to open and shut in the same way.

Why do you suppose they were given so queer a name? Perhaps their colors reminded some one of eggs that have been scrambled with butter in a hot pan.

The plant has other odd names and is called yellow toadflax, brideweed, and impudent lawyer. It has still another name. A Welsh gentleman whose name was Ranstead is said to have introduced this plant in Philadelphia gardens years ago. So, when you learn that one of its various names is ranstead, you will understand that this is in honor of a man who considered the plant good enough for flower gardens. It may be that you will agree with the Welsh gentleman.

Butter and eggs is much more likely to be found listed in a book of weeds than in a seed catalog of a florist even though it is much prettier than many yellow flowers that are sold for rather high prices.

Originally butter and eggs grew wild in Europe and in parts of Asia. Its seeds often become mixed with the seeds of field plants grown for crops, as are the seeds of many other weeds. In this way they have been carried from place to place as uninvited guests. They have made themselves at home along gravelly sides of railroad tracks and highways, as well as in other dry, sunny places.

If you like yellow flowers, you may wish to have butter and eggs in your own garden. Then you can have a pleasant time watching a bumblebee open the mouth of one of the flowers. The fuzzy insect is heavy enough to cause the mouth to gap open while she rests on the lower lip. She thus has a chance to poke in her head and dip the tip of her long, stiff tongue into the nectar. While she is taking her drink she becomes dusted with pollen, which she carries to the next flower she enters.

Butter and eggs and snapdragons are called "bumblebee flowers" because they depend on these insects to carry their pollen. Bumblebees like to drink from these flowers. They welcome the yellow butter and eggs even when people do not.

Mullein

The great mullein, or common mullein as it is also called, is another native of Europe. Its seeds, with those of many other weeds, have been gathered with those of crop plants and thus carried to different countries.

Mullein seeds are fine and light and are blown like dust by the wind. Plants growing from such scattered seeds may now be found along roadsides and in fields all the way from Maine to California. Although it is not a native, mullein has become naturalized in this country and is one of our common plants.

Most people look upon the mullein as a weed, and you may wonder why it is given a place in this chapter. It really has an interesting history as a garden plant.

One day an American florist looked thoughtfully at a young mullein. The soft, thick leaves of the rosette it forms during its first season seemed attractive to him.   He decided that "velvet plant" would be a good name for mullein.   So he listed it with his

Rosette of common mullein, or "velvet plant"

other garden plants and offered it for sale. For some years after that, people in America and Europe, too, purchased "American velvet plants" and gave them honored places in their gardens.

The mullein is indeed interesting, from the months of its fuzzy, overwintering rosette to the time of its full height the next season. You are likely to see this tall and stately plant growing beside the road when you are riding in the country. Its spike of golden blossoms may make you think of a flaming torch.

You may like to learn that many years ago the Romans dipped dry mullein stalks in tallow and burned them for torches in their parades.

In ancient times the Greeks, too, had a use for mulleins. They dried the leaves and placed them in their open oil lamps for wicks.

If you put a mullein into your garden, you may enjoy it more than some of your other plants. Perhaps you will see a humming bird come to gather fuzz from its leaves to line its dainty nest. Perhaps you will see a flock of twittering goldfinches come to feast

on the seeds its seed pods hold. Perhaps —
but you will not care to have us tell you too

A tall and stately mullein

much about the mullein. You will wish to
discover for yourself some of its interests.

Turtlehead

We have told you of a few members of the Figwort Family that have been introduced in America from Europe. Now we shall speak of one of their American relatives — the turtlehead, which is also known by the names of snakehead, balmony, and shell flower.

You are likely to meet this plant in such moist places as swamps or ditches or the borders of streams. You would not be wise to move it into your garden unless you have a good wet place to put it. But you will be interested to visit it in places where it grows naturally. It may be found all the way from Newfoundland on the north to Florida on the south and westward as far as Alabama, Kansas, and Manitoba.

You may like to notice in what ways the white or slightly pinkish turtlehead flowers resemble, in shape, the snapdragons and other figworts we have mentioned.

QUESTIONS AND ACTIVITIES

Have you ever seen a turtlehead? Look for a picture of one in some book of wild flowers. (Then if it is convenient for you to do so, hunt in a swampy place for some of the plants.)

How does a bumblebee get into the mouth of a turtlehead? Why does the beautiful Baltimore butterfly come to this plant to put her eggs on its leaves? After seeing her come and go, can you guess what caterpillars enjoy a salad of turtlehead leaves?

Make a list of some plants you would like to find growing wild in country places. Make a list of plants you think you might find in a city park. Perhaps you would like to keep these lists until some pleasant summer day when you may have a chance to visit a country field or a city park. Then you can try to find the plants that are named in your lists.

The chapter you have just read is about a few plants belonging to one family. Choose some other plant family. Write a short essay of your own about a few plants belonging to the family you choose. Write your essay about plants of this family that you would like to have in a flower garden. Here are some suggestions:

Rose Family — rose, Japanese quince or fire-bush, bridal wreath, flowering raspberry

Lily Family — day lily, lily of the valley, tiger lily, Turk's-cap lily

Composite Family — aster, dahlia, daisy, marigold, sunflower

If you choose plants of the Rose Family for your essay, tell how many petals a wild rose blossom has. If you write your essay about plants of the Lily Family, tell how many divisions there are in the funnel-shaped outer part of the flower of a tiger lily. If you do not have a wild rose or a tiger lily to study, find a picture of each in some flower book.

Does the turtlehead grow wild in your state? If you do not know, look at a map to see if the state in which you live lies somewhere between Newfoundland and Florida and not farther west than are Alabama and Kansas.

Read pages 3 to 16 again and make a list of all the plants mentioned that have the names of animals as part of their own fanciful names. *Toad-flax*, of course, would be one such name.

GARDENS FOR GUESTS

Many houses have guest rooms — rooms which are not commonly used by members of the family but which are reserved for visitors.

Perhaps, however, there is no part of a home where our guests may be happier than in a garden. If you plan a garden, you may like to have one to which you invite your friends in the cordial manner of Thomas Davis :

Come in the evening, or come in the morning;
Come when you're looked for, or come without warning.

The boys and girls you invite to your garden may read or play there in good summer weather as cheerfully as they could in an indoor room. Gardens are pleasant places for parties in the afternoon with something delicious to eat about four o'clock. Evening parties, too, are jolly, for then the gardens may be made gay with Chinese lanterns.

But of course people will not be the only guests that you invite to come. No garden is perfect without certain insects and birds.

Lilacs for Butterflies

When you invite people, you do so with words — written words in notes or spoken words. When you invite butterflies, you use color and fragrance.

Do you wish to see swallowtail butterflies, frail black and yellow creatures, flitting and floating in the sunshine? Then give them a lilac invitation!

Any early lilac will do, for it will blossom about the time these butterflies leave the chrysalis cases in which they have spent the winter. It is not necessary to buy some ex-

Lilacs for butterflies and clearwings

Some will be honeybees and some will be bumblebees. Their wings are not silent like those of the butterflies. They make a pleasant murmur in your garden.

There are other guests, too, besides the butterflies and bees. Watch for the clearwing moths. You may easily mistake them for bumblebees. Many people do not notice the difference. See if your eyes are keen enough to discover which fuzzy black and yellow bodies belong to bumblebees and which are those of the clearwing moths. It is easy to guess why one name for this kind of insect is "bumblebee moth." Listen to the whirring of their wings. Can your ears detect the difference in the tune of the bumblebee and that of the clearwing moth?

Clearwing moths have larger and stronger wings than those of bumblebees, and their flight motions are different.

pensive variety.   The commonest kinds
the butterflies just as well as the rarest.
haps they will suit you as well, too.

Watch your lilac bush when it blosso
Do you love the color in the sunshine
the fragrance that drifts with the air?   ʼ
are not alone in your joy, for here come
butterflies!

The scent reaches them while they are s
far off.   They accept your invitation a
come to the garden party.   You need n
offer them lemonade with straws in glasse
The lilac bush provides the treats.   Eve
blossom has a slender cup with nectar tha
tastes better than lemonade to butterflie
Their own long tongues can be uncoiled an
used to suck the sweet liquid.

Lilacs for Clearwings

Butterflies are not the only insects that
accept the lilac invitation.   Bees will come.

Petunias for Humming-Bird Moths

The clearwing moths belong to the family called Hawk Moths. These insects were given this name because many of them have long, narrow wings and are strong in flight. Another name for this same family is Humming-Bird Moths. Some of them resemble the humming birds in size and shape. They hover before a flower in much the same way as humming birds do. It is only one group of this family that have clear (transparent) wings. All the others have the wings entirely covered with colored scales.

We have remarked that some people never notice the difference between a bumblebee and a clearwing bumblebee moth. Many people (perhaps the same persons) never know whether they are watching a humming bird or a large hawk moth.

A few kinds of humming-bird moths fly

during the day, but most of them wait until twilight. They like moonlight hours, too.

You may have read that they visit evening primroses. They visit other night flowers, as well; and they are very quick to accept invitations from petunias.

Do you like petunias or do you think of them as rather floppy, sticky, stale-smelling plants that you do not choose to have in your garden? If you have never learned to appreciate petunias, it may be because you have not become really acquainted with them.

Our advice is to give them at least a corner in your garden — an out-of-the-way corner if you prefer. Then visit them at twilight, the twilight of evening or of early morning as you choose. And if your family do not object, go into the garden some summer night when the moon is bright and full.

What is that fragrance that greets you as

Petunias for humming-bird moths

you walk down the garden path? It seems to come in waves as with a gentle wind. There is nothing stale about it now. The odor is amazingly fresh and sweet and spicy. Your petunias are inviting the humming-bird moths to come.

You may be surprised when you stand beside these plants. Their blossoms show more plainly in the moonlight than you might expect. They are almost shiny in the night — *luminous* is a better word for them. This is when they are at their best; and now is the time when those night moths come to poise, on whirring wings, like humming birds before each vaselike flower.

If you hear anyone say, "I do not like petunias," you may know that that person never visited them in a garden at twilight when he was a child. He has no memory of an hour when marvelous little creatures of the night came whirring through the waves of spicy fragrance and stayed to sip from the flaring cups these gracious plants held up for them.

Pollen and Nectar for Bees

Butterflies and moths do not eat pollen. It is only the nectar they seek in flowers. To be sure, each of these insects usually gets its head dusted with pollen grains from the stamens of the flower while it is drinking nectar there. Some of this pollen shakes off into the next flower it visits and clings to the sticky top of the pistil there. Later these pollen grains reach down into the seed case at the base of the pistil and help form the live seeds.

In this way the butterfly and moth guests pay for their sweet drinks although of course they do not know they are helping the flowers. They merely feel thirsty and drink because it is pleasant for them to do so.

Some other kinds of insects, however, visit flowers only for pollen. Some flies do this. So do some beetles. Such pollen eaters do not rob the plants. They do not eat all the pollen. Enough of it sticks to their bodies so that they carry it from flower to flower as moths and butterflies do.

Bees seek both pollen and nectar, as you doubtless know. Many garden flowers are visited by bees. We think you will not need to be told what some of these flowers are. It seems quite likely that you have watched honeybees and bumblebees so often that you could write an essay of your own about these bees and blossoms.

Leaf-Cutter Bee

One of the most interesting of the many bees that come to the garden is the leaf cutter. She comes for nectar. She comes for pollen. She comes, too, for pieces of tender green

leaves.   Her jaws are her scissors, and she cuts
the pieces in neat circles.   But she does not
eat them; she takes them away and uses
them in building her thimble-shaped nests
in some old post or dry, hollow stem.

Photo by Cornelia Clarke

**A leaf-cutter bee has cut pieces from these rose
leaflets.**

We have heard a few people complain
about leaf-cutter bees.   They said they did
not like to have any little circles cut from
their plants.   We think, however, that it is
a rather stingy person who does not welcome
these bees to his garden.   They do their full

share there as pollen bearers. Do they not earn the few little circles which they take without injuring the health of the plants?

You may find the plants from which the leaf cutters have taken their building materials quite as interesting as those that have not been touched.

Red Flowers for Humming Birds

Humming birds come to most kinds of flowers that have short or long nectar tubes. They drink from the tubes of white, blue, and yellow blossoms. They will also come to apple blossoms, geraniums, hollyhocks, and many other flowers that hold nectar in shallow parts instead of in tubes. They take tiny insects for food besides the nectar that they drink.

Red is the color, though, that attracts them most of all. Perhaps they can see this color farther than others. For some reason

the best invitation to bring humming birds
is a red one. So, if you would welcome these
little guests to your garden (and who would
not?), have plenty of red flowers for them.

Which is the humming bird?

Scarlet runner beans are good to plant if
you have a place for them to climb. If there
is a spot in your garden that is wet enough
for cardinal flowers, you will like to have
those, we think ; and so will the humming

birds. They also visit tawny red day lilies, wild columbines, vivid bee balm, and — but we need not give you a long list. Any red blossoms with nectar are excellent for humming-bird flowers.

These birds are, indeed, so attracted by red that they often make mistakes. They will fly to a bush covered with red berries as if expecting a treat and then stop, in a puzzled way, and turn aside. They will fly to a bright red porch cushion and hover before it until they discover that it has no refreshment to offer them.

We repeat that humming birds will come to flowers of any color and of many shapes. But if you wish to make them particularly welcome, try red for a color and tubes for shapes. Try, for example, a trumpet creeper or a trumpet honeysuckle. Or, if you prefer, try the scarlet blossoms of Oswego tea, also known as Indian's plume.

Berries for Birds

Did you ever see a fat young robin, in its first suit of spotted feathers, sitting on a twin-flowered honeysuckle bush, begging for food?

He teases while his mother picks her bill full of honeysuckle berries, which she spills into his greedy mouth. Then if she flies away, what do you think that young rascal does? He sidles carefully along the branch and picks a few berries for himself.

It is fun to watch the berry eaters feast in a flower garden. They have such jolly times from sunup to sundown that it is a pleasure to have them for guests.

A berry bush is enjoyable in three ways — for its beauty of blossoms and berries and birds.

Some bushes bear fruit that is ripe for birds in summer, such as the twin-flowered honeysuckle and elderberry bushes. There are a number of cornel bushes (also called *Cornus* and dogwood) that have berries birds like during the fall days. Some kinds of cornel bushes have red berries, others have blue fruit, and still others have white. High-bush cranberries hold their fruit until spring if it is not gathered. So, if you have room for a clump of these, you may give a winter garden party to pine grosbeaks, evening grosbeaks, and other hungry berry-eating birds that are about during the cold months.

QUESTIONS AND ACTIVITIES

Write a short essay of your own about "Flower Gardens for Bumblebees."

If *Surprises* and *Through Four Seasons* * are where it is convenient for you to see them, look at the pictures in these books. Make a list of the bee flowers (flowers that bees will visit) that are pictured in these books.

* References to certain books will be made from time to time. See page 424 for a list of the books that are mentioned.

Explain how flowers help bees.   Explain how bees help flowers.

Read "A Tuft of Evening Primroses" in *Holiday Hill*.

Write a short essay about a humming-bird moth.

Bumblebee and clearwing moth

Draw a plan for a garden with red flowers for humming birds. Make a mark (x) for each plant you wish to have in your garden.

Make a list of all the red flowers the names of which you can easily remember.

On the previous page is a picture of a bumble-bee and a clearwing moth. How can you tell from this picture which is which? The mouth parts do not show in the picture. One holds its sucking mouth parts flat against the under side of its body when it is in flight. The other holds them coiled like a watch spring, as a butterfly does, when not feeding. Try to find out whether the bee or the moth coils its mouth parts.

HOME · SHELTERS

PART TWO

A Canadian Indian wigwam

DIFFERENT KINDS OF WALLS

Any shelter that shuts out cold winds must have walls or sides. Men have used different materials for the sides of their homes.

Shelters of Hides

Skins of animals have been dried and used for this purpose. A great many years ago Greek and Roman and other European soldiers had small tents for shelter. Some of these tents were made of skins (hides) of animals. In many parts of North America Indians once lived in tents or lodges. Some of these homes were called wigwams. They were often made of animal skins. Eskimos have tents of skins for their summer houses. Tents of hide have been used, too, in other parts of the world besides Europe and North America.

41

Shelters of Snow and Ice

Children living in the northern parts of the United States like to build huts with walls of snow in winter. White men in lands of cold

Courtesy V. S. Jonsson

Building an igloo of snow

winters sometimes build palaces with walls of gleaming ice. Such huts and palaces are not to live in, of course. They are made for fun.

Farther north, however, in the arctic regions, snow houses are really used by some Eskimos in winter.

In most places Eskimos live during the winter in houses of sod (earth) with supports of wood, stone, or bone. In some localities, as you have just read, they build houses of snow for winter use. In still other regions these northern people have both sorts of dwellings — the earth houses for their regular homes and snow houses for temporary shelter when they are away on hunting trips. The Eskimo word for *building* is *igloo*.

Hides or snow, of course, are useful only in a few sorts of simple buildings. Most people in the world have shelters that are different from these.

Shelters of Plant Materials

Many homes have walls made of materials taken from trees or other plants. Skilled carpenters are needed to do the woodwork in modern houses. A whole chapter (pages 81–101) is given to carpenters and their work.

Courtesy V. Stefansson

An igloo of logs

So we shall mention here only some of the shelters that are easy to build with parts of plants.

Grasses furnish materials for homes for

certain creatures. Many birds gather dead grass stalks for the walls of their nests. They take them when they are soaked with dew or rain. The stalks are pliable while they are moist, and they can then be handled without breaking. You may have noticed that birds work at such nests early in the morning instead of during the middle of a sunny day.

Field mice and some other rodents use dead grass for their nests.

When a bumblebee wakens from her winter sleep and flies off on her house-hunting trip in spring, she is most likely to move into a grass home that a field mouse has built and used.

Men are not slower than birds and mice and bumblebees to appreciate grasses as building materials. They have often used the stalks of wheat or other straw to thatch their roofs. But the grasses men have found most useful for walls are the bamboos.

Some bamboos grow more than a hundred feet tall, and their stalks may be a foot through. They often grow in forests like trees. They are the giant members of the Grass Family.

The stems of grasses are strong, round tubes. The tubes of some are quite hollow, as you know if you have ever sucked lemonade through a real straw. Corn and some other kinds of grasses have light *pith*, or spongy matter, inside their tube-shaped stalks.

People who have studied objects of different shapes have learned that they are especially strong if they are built in hollow, circular tubes. A hollow, tubelike post of steel is stronger than a solid steel post that has the same weight. A rubber hose would not be so strong if it were a square instead of a circular tube.

The hollow in a grass stem is not an open tube all the way from the ground to the tip.

There are knobby places that close the opening like a knot.   One name for such a knot
is *node*.   The long, flat leaves are called
blades because they are shaped like sword

Courtesy U. S. Dept. of Agriculture

There are knots, or nodes, in cornstalks.

blades. A grass leaf does not grow on a
slender stem, as many leaves do. It is
wrapped about the stalk at its base. This
wrapper part of the leaf is called the *sheath*.

One reason the stalks of grasses are strong and rigid when they dry is that they have a great deal of *silicon* in them. Silicon is more abundant in the earth than any other *element* except oxygen. An element is a simple substance. That is, an element is not composed of two or more other substances, as *compound* substances are. Silicon is never found by itself, as it always combines with some other element. Sand and certain rocks are largely silicon. Men add it to cast iron of special kinds.

Grasses take silicon they find in the soil and add it to their stalks. It helps make them strong. Bamboos use a great deal of silicon. Some bamboos have so much of this substance that sparks fly when their stalks are chopped with a hatchet. The same thing happens when a flint stone is struck with iron.

Men have found many uses for straight,

strong, light bamboo stalks. They make canes and fishpoles and furniture with them.

Photo from R. I. Nesmith and Associates

The doorway of a bamboo house in the Philippine Islands

These giant plants are most abundant in warm climates. Bamboo houses are common in tropical countries.

Shelters of Earth Materials

Rocks, clay, and other earth materials have long been used by men for the walls of their homes.

Trained workmen are needed to make many of these homes. They are the masons about whom you will read in the next chapter.

The very simplest of all rock houses can be used without any help from masons to make them ready. These are caves in which shelter may be found.

A long time ago (thousands and thousands of years ago) certain European bears and lions and hyenas used caves for homes. Bones of such animals have been found in the caves. They are not of just the same sizes and shapes as those of any bears or lions or hyenas that are living today. So we know they were of different kinds. We call them cave bears, cave lions, and cave hyenas.

About the same time there were men in Europe who lived in caves. We call these ancient people cave men. Some of their skulls and other bones have been found.

Cave men doubtless sought their caves as naturally as the bears and other cave dwellers did — to find shelter from bad weather and from their enemies. It did not take a great deal of thought or much intelligence to seek and use such places.

It took human intelligence, however, to do some of the other things that the cave men did. They made knives and scrapers of stone and other sorts of stone tools. Their tools have been found with their bones in the caves.

These ancient men had the thoughts and skill of artists. They left pictures on the walls of their caves — pictures on the rocks that have lasted all these thousands of years.

In North America, certain early Indians made their homes on shelves of rock canyons

Where ancient cliff dwellers lived

in the Southwest. They found protection there from other Indian tribes. We speak of them as "cliff dwellers." They made beautiful pottery; so do Indians living in that part of the country today. Some of the old pottery has been discovered in long-deserted cliff dwellings.

If you found yourself in an unsettled place where there were no caves or trees, what would you do for shelter? Dig a hole in the ground, perhaps? That is what many a pioneer did in the early days on the western prairies. He called such a place a *dugout* and used it until he had time to make a better sort of home. He and his family could keep snug and warm in a dugout — as the families of gophers and moles and many other creatures do. But, of course, the pioneer made a different sort of home as soon as he could.

An easy kind of house to build is a sod house. Anyone who can build with blocks can make walls with blocks of sod. Sod houses may be quite comfortable homes. They were common prairie dwellings in the West during pioneer days. They are still

Courtesy William Hawk

A neat sod house with a shingled roof

used in some places far from lumber or quarries. As you have already learned, many Eskimos build sod igloos. The old sod houses had rough roofs, but today some of them have good shingled roofs.

Rammed-Earth Walls

Some buildings have walls made of hard, packed loam. They are called *rammed-earth* buildings.

It is known that rammed earth was used in some parts of the world many centuries ago. A general named Hannibal had walls of rammed earth built for his soldiers in Spain and Africa more than 2000 years ago.

Rammed-earth houses are common today in a province of France where it has long been the custom to build them. There are some houses of this sort in England and in the United States.

Most varieties of earth are suitable for ramming. Pure clay is not good for this purpose because it shrinks too much in drying. Pure sand will not do because it does not hold together (bind) well. But a mixture of sand and clay is satisfactory. So are many other ordinary kinds of earth.

Forms to hold the earth while it is being rammed are needed in building walls of this sort.

Courtesy H. B. Humphrey

Mold (or form) in which earth is rammed

The house which is shown in the picture on the next page has a tile roof that weighs eighteen tons. The rammed-earth walls of this house are so strong that they could support more than one hundred times that weight.

Does it seem strange to you that walls made of nothing whatever but ordinary earth

Courtesy H. D. Humphrey

A house with rammed-earth walls and tile roof

should be so strong? It is the pressure of the ramming that packs this earth into a hard mass — so hard, indeed, that it can no longer be dug with a spade. It must be cut with a chisel as rock is, if it is necessary to make a hole in it. Rammed earth is a sort of man-made rock.

In another chapter (page 72) you will learn that pressed brick is denser and more rock-like than brick that has not been pressed. Some kinds of rocks that are dug from the ground were made from small particles that were packed and pressed into layers of solid rock. Sand and mud and bits of shells sank to the bottoms of oceans and lakes. The top layers of such materials were heavy weights pressing down on what was beneath them. We might almost speak of such rocks as "natural rammed earth."

QUESTIONS AND ACTIVITIES

You have read about houses of :
(1) Hides
(2) Snow or ice
(3) Wood or other plant materials
(4) Rocks or other earth materials

Can you think of any material for a home-shelter that would not be one of the four sorts in this list?

Which one would a canvas tent be? Which one would a brick house be?

If you care to do so, you can make a toy wigwam with material cut from an old glove of kid or some other kind of leather. Would that material be No. 1 or No. 3 in the list?

A hollow, round straw is strong for an object that weighs so little. The quill of a bird's feather is a hollow, round tube. Would it be stronger and lighter if it had some other shape?

Make a list of large or small hollow tubes that people use.

Look at a stalk of some plant of the Grass Family. (A bamboo rod, cornstalk, or stem of growing or dry grass or any other plant of this family will do.) Find the solid knots, or nodes, in the stalk.

Would you like to make a small block of rammed earth? See how firm a mass you can make by ramming some slightly moist earth in a tin cup. Use the end of a pencil for a rammer.

Set a wooden pail in a hole in the ground so that

it will not move. Ram some earth in the pail until it is hard. If you cannot dump the rammed earth from the pail the first day, let it dry a while before you try again to get it out. Can you dig into it with a trowel? Can you break it with a hammer?

THE MASON'S WORK

A mason builds with stones and bricks. He uses mortar and cement. He works with earth materials.

Ages before men learned to be masons, there were other creatures that could build with clay or stones.

Bird and Insect Masons

Cliff swallows made homes of clay centuries and centuries ago, just as they do today. They gathered the clay in their bills. There were usually bits of dead grass or broken roots or similar fibers in the clay they found. Such fibers helped hold the clay together. The heat of the sun baked these little clay buildings until the walls were hard and strong enough for the birds to use.

Cliff swallows and clay nests

Long, long ago, there were certain kinds of caddis worms (larvae, or young, of caddis flies) that made shelters for themselves with pebbles, as they do still. They found the

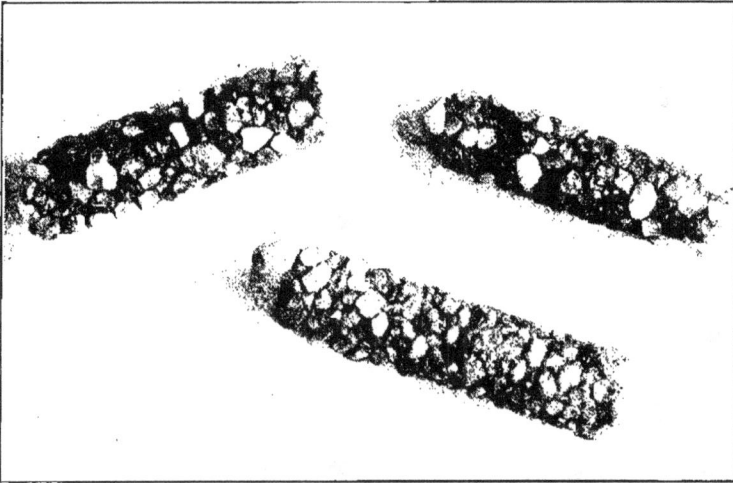

Stone buildings made by caddis worms

pebbles in the water where they lived. Each caddis worm made its own little tube-shaped rock house. It fastened the tiny stones together with silk that oozed from its silk glands. One end of the tube was open and the larva could push its head and legs out or pull them inside. It could drag its little stone house from place to place as easily as a snail moves its shell.

Even in those far-off years, too, before men learned to be masons, there were mason

Mason wasps and clay buildings

wasps. They made jug-shaped homes of clay for their young. They carried the clay in their mouths and mixed it with a sticky sort of saliva. This saliva was their own kind of cement for making their clay buildings firm and strong.

The ancient cliff swallows used their bills for tools. The caddis worms and mason wasps used their jaws and front feet. They were natural masons working with natural tools.

Men, however, could not be masons until they experimented and learned how. They began to make rough stone tools with which they could break rocks. They found ways to make their blocks smooth and to fit them together in buildings.

Buildings of Stone

For many years stone houses seemed too imposing for any people, except powerful rulers, to have for homes. Buildings of stone, indeed, were so impressive that they were considered suitable places of worship. So stones were used mostly for temples and monuments.

Perhaps you have read of the huge stone pyramids of Egypt and looked at pictures of them. Old temples made in Greece and Rome were so stately that people still take great care of their ruins. Indeed, even yet, when men wish to make beautiful and dignified buildings, they often use Greek temples for models.

It is not necessary, however, to go to Africa or Europe to see grand old temples of masonry. Remarkable stonework may also be found in Mexico.

Stones are hard enough to use for building just as they are taken from the ground. Workmen may break or cut them to required sizes. They may smooth or even polish the surfaces. But they do not change the stones themselves.

Different sorts of stones with many special names are used in buildings. The principal building stones are the granites, which are

very hard and dense; the sandstones, which
are less hard; the limestones, which are the

Courtesy Carnegie Institution of Washington

An ancient temple in Mexico

least hard of the stones commonly used for
building purposes; and the marbles. The

marbles were once really limestones. They became harder because of heat and pressure within the earth itself at some very ancient time.

Cave men and cliff dwellers and other primitive men needed to live in caves and cliffs and quarries in order to have stone walls about them. We do not need to live where the stones are. We can bring the stones to the places where we wish our homes to be.

In some places it is not necessary to carry stones far to use them. People living in the mountains of north Wales, for example, can find stones as easily as the Pilgrims could obtain logs in New England. Stones are as near them as sod is to people living on a prairie. Stone houses in such places do not cost much.

Even large marble buildings would not be so expensive as they are if they could be put up near the quarries from which the stones are taken. But of course it costs a great deal to carry stones a long way by steamer, by train, or by automobile truck.

It was easy to find stones for these houses on the mountains of north Wales.

In spite of the cost, however, stones may be worth bringing long distances. Some are so durable that they last for hundreds of years.

They withstand wind, rain, heat, and cold much better than most building materials. Also, they can hold enormous weights and stand great pressure.

Buildings of Brick

When the mason comes to build a house, he may not use stone at all. He may use another earthy material. Perhaps he will use blocks of clay that have been baked. We call such blocks bricks.

Bricks may be baked in the sun. Many of the bricks used in ancient Egypt were sun-baked. They were made of clay mixed with straw.

Sun-dried bricks are used in regions of southwestern North America where there is little rainfall. They are called *adobe* bricks.

Bricks baked by the heat of the sun are not durable in places where there is much rain.

In most parts of the country the clay blocks are baked in ovens, called *kilns*, until they become hard.

Courtesy "Better Homes in America"

An adobe building

The minerals in the clay determine the color of the brick. Because of this, men sometimes mix different kinds of clay to get a particular color of brick. This color is not always uniform. It may appear as uneven spots on the surface of the brick.

Men make a kind of brick by pressing the clay blocks hard with a machine. After such

Courtesy Harbison-Walker Refractories Company

A brick kiln

bricks are baked, they are very dense and strong. They are called *pressed* bricks.

Sometimes bricks are cut in a way to

roughen the surface. Of course such bricks reflect less light than bricks that have a smooth surface. This gives them a different, duller look.

Terra cotta is another kind of material made from clay. It is often molded in special forms and made to look like cut or carved stone. It may be colored. It is often used to decorate buildings made of plainer materials. *Terra cotta* means "baked (cooked) earth."

Mortar and Concrete

Whether the mason uses stone or ordinary brick or terra cotta, he usually needs some substance to hold together the many pieces that he puts into the walls of a building. He uses *mortar* for this purpose. Mortar may be plaster or cement.

Not many years ago people made most of their plaster by mixing sand, lime, and hair taken from the skins of cattle or other animals. The hair fibers help to hold together, or bind, such plaster.

Now a great deal of plaster is made with *gypsum*. Gypsum is a mineral. It is taken to factories to be made suitable for mortar.

Cement was sometimes used between stones in ancient masonry. Modern masons still use cements that they call by old Roman names though they make them in a different way now.

Millions of barrels of cement are made every year. The materials used for cements are clay, shale, slate waste, blast-furnace slag, limestone, chalk, and marl, although not all these are put into any one kind of cement. Marl or shale and clay, for example, are used for one kind.

After the materials for the cement have been selected, they are ground to a fine, fine powder. It is possible to mix things more thoroughly if they are in the form of powder than if they are in larger particles. The mixed powder is burned in a great kiln.

When the burned material — the cement — is cool, it is crushed to a powder so fine that most of it will pass through a sieve which has 40,000 openings to the square inch. The cement is then put into bags or barrels, ready to be sent where the masons need it for buildings.

Cement may be used for mortar by mixing it with water and not too much sand to make a paste that can be spread evenly. As it dries it becomes hard.

When men make *concrete*, they mix more sand with the cement and water and they add pieces of broken rock. The foundations of most modern buildings are made of concrete. Sometimes the basement walls, only, are of this material. Often the walls of a whole building are of concrete.

Workmen pour the wet, freshly mixed concrete into molds, or forms. When it dries, it is like great hard rocks of just the right shape to serve as sections of the building.

Another craftsman who may come to help make our houses works with some of the same materials that the mason uses. We call him the plasterer. He puts his plaster over the laths if the carpenter has nailed any on the walls. Some of the wet plaster is pushed between the laths and clings to their edges. The laths thus serve to help hold the plaster in place after it has dried.

Often, however, laths and plaster are not used. Many builders use plaster board instead. This is really a sort of plaster made in the form of sheets. A carpenter can nail and saw plaster board as he does lumber.

The Mason's Tools

When the mason comes to work, he brings special tools, for he must often cut and shape his building materials. His hammers must be of different shapes and very strong. He learns how to strike a brick to have it break in the way he wishes, and he must know how

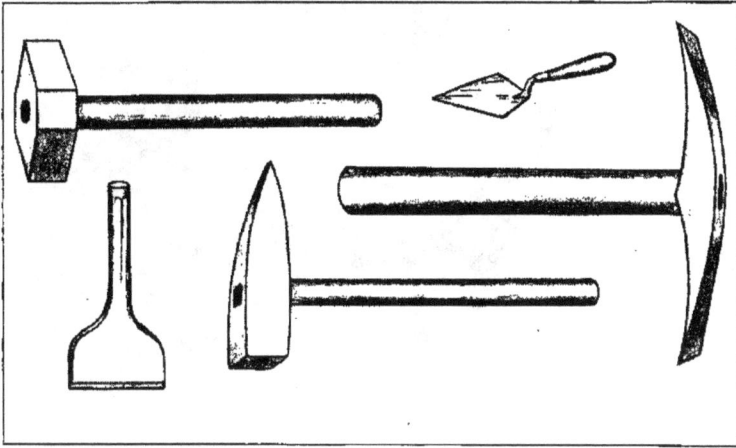

Tools used by the mason
What is the name of each tool?

to break stones along the lines where they split easily. He needs chisels, too, for the finer work. He may even have saws made especially for stonework.

Trowels are necessary tools of the mason, also — a big one for buttering great stones

with mortar and a small one with a sharp
point if he wishes to work a design in the cement before it hardens.　Brushes are useful,
too, either when the surface is being cleaned

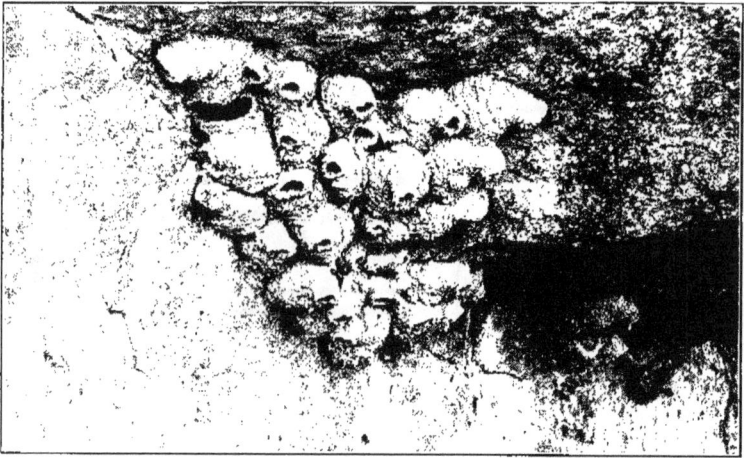

Courtesy U. S. Bureau of Biological Survey

**The clay nests of cliff swallows are the same kind
today as they were thousands of years ago.**

or when it is being worked to make it rough
instead of smooth.

For hundreds and hundreds of years human
masons have been learning their trade.　They
have taught one another.　They have discovered different ways to use their materials.

They have been glad to accept newly invented tools and machinery.

But mason wasps still make the same sort of cement for their homes that they made in the long-ago age of the cave men. Caddis worms make their stone houses after their own ancient models. And cliff swallows still build their nests of clay in the same manner as they did in the days of the cliff dwellers.

QUESTIONS AND ACTIVITIES

The walls of a cliff swallow's nest resemble adobe more than they do pressed brick. Explain why this is so.

If you can get some clay to use, try to make a good sun-baked brick.

Are workmen making the foundations of a building near enough for you to watch them? If so, find out whether they are making the foundation walls of brick or stone or concrete. Write a short account of what you see the workmen do.

Do either No. 1 or No. 2, as you choose:

(1) Go to look at some old house. Write a

short description of the foundation wall. Tell how you think it was made.

(2) Go to look at the newest building near your home or school. Write a short description of the foundation wall. Tell how you think it was made.

Would you like to see pictures of ruins of temples and other masonry made many years ago? Then look for them in books about ancient Babylon or Rome or Greece or Mexico. Some histories and books of travel have such pictures. Your teacher or a librarian may be willing to suggest books that will be convenient for you to use.

WOOD AND THE CARPENTER

A carpenter is a builder of wooden structures. In order to understand his work, it is necessary to know something about the materials he uses.

Any boy or girl who has sharpened a pencil with a knife has learned that wood is easy to cut. Of course only a person with proper tools can work well with wood. Tools are, indeed, as necessary to a carpenter as they are to a mason. For this reason the first creatures to use wood for homes were those that had some parts of their bodies suitable for tool-like uses.

Bird and Insect Carpenters

Woodpeckers cut neat and cozy homes in trees untold centuries before any man ever

had a chisel. Wood-boring beetles made

Photo by Lynwood M. Chace

A woodpecker uses his bill for a tool.

their even tunnels through branches and
trunks of trees ages before man invented the

first gimlet or bit. Carpenter ants and carpenter bees and carpenter moths made themselves at home in comfortable tenements of wood long, long before a human being constructed his first house of this material.

Caterpillars built these log cabins.

Some creatures, to be sure, that use wood for their shelters do not hollow out cavities for the purpose but make little log cabins of sticks laid with their ends crossed. Many birds make their nests with such foundations. And there are caterpillars of a certain kind, or species, known as Abbot's bagworm, that cut twigs and fashion them into most attractive buildings.

Log Cabins

Men, too, have followed similar methods of placing sticks and even trunks of trees so that their ends are crossed in such ways as to help hold one another in place.

Neat and strong log cabins can be made with very few tools, the most important of which is an ax. They have been the easiest houses for early settlers to build in many parts of our country. The people in colonial days found plenty of trees in near-by forests. They cut logs from the heavy trunks to make thick walls. They filled the crevices with clay or other plaster and had snug shelters.

Such crude log buildings really looked

much better in their woodland setting than more elaborate houses. People still like to see log houses in such places even when they

Courtesy "Better Homes in America"

Nine boys are working on this log cabin.

might build other kinds. So it happens that lumber camps are not the only places where we see log cabins today. Log cottages are to be found in many fashionable

summer camps.   Large modern hotels made
of logs may be visited in our national parks.

Preparing and Using Lumber

In most places, however, log cabins are
no longer built.   Carpenters, now, do not
usually work with whole logs.   They need,
instead, lumber cut in regular shapes and
sizes.

Woods of various kinds of trees have great
differences in grain, strength, color, and
other qualities.   Some wood (such as balsa)
is so soft that you could press a nail into it
with your hand.   Other woods are so hard
that the same nail would bend or break if
you tried to pound it in with a hammer.

Wood from some trees will stand great
weights, while wood from other trees is not
suitable for places where strength is needed.
Some kinds that are useful for rough rafters
do not have a beautiful grain.   So they could

not be made attractive for places where un-painted wood is desired for inside finishing or decoration.

Courtesy U. S. Forest Service

Logs in a river

Trees must be cut in the forest and taken to sawmills, where lumber is prepared for the carpenter to use.

Perhaps you have seen one of the small sawmills to which tree trunks are brought on wagons or trucks. You may have visited one of the great sawmills to which the logs come by trainloads or by floating down rivers. Many of the biggest mills have been put where the logs can be brought by rivers. The trees in the great forests are so enormous and heavy that men choose the easiest way to handle them. In the North men go into the woods in the winter, when the snow makes it easier to pull the great logs to the river's edge. In spring, when the ice melts, the rush of water carries the trees down until they reach the mill.

A great sawmill is an interesting place. It receives the rough logs and changes them to lumber properly shaped for the carpenter's use.

The first mills were equipped to do no more than saw the logs into boards. The carpenters did the rest of the work while they were constructing the buildings. They

Carpenters use lumber that comes from the mills.

made the doors and window sashes and window frames. They even made the shingles with hand tools.

But men have invented machines that can do many kinds of woodwork much more

rapidly, and sometimes better, than carpenters can with hand tools. Shelves and
cabinets and many other parts come almost
ready to be put in place. Indeed, in recent
years, whole houses of certain kinds can be
bought ready-cut and fitted, so that all
the carpenters need to do is to put the parts
together.

If you watch carpenters build a house, you
will see them put the frame quickly in place.
They put up the big timbers first. Then,
as the structure grows, they fit smaller parts
into place.

Carpenters begin with great spikes for the
heavy timbers. They use slender nails with
small heads for the inside woodwork. They
have other nails, too, of many different
shapes and sizes for the various places where
nails are needed.

But, in spite of all these useful nails, some
carpenters still like to fit wooden pegs to hold

framework, and even floor boards, in place. This is done only where very careful work is required and the time spent in using hand-made parts is considered worth while. For some places a peg that can be whittled into shape with a jackknife is preferred to the best machine-fitted nail. But, of course, that does not mean that any carpenter would wish for the old days when there were no metal nails at all. Nor would he wish for that time, not so long ago, when iron nails were made only by hand at a forge and were very expensive.

Building Boards

As you know, the chief material of the carpenter is wood. Even the heavy paper he uses is made from wood pulp. He puts sheets of this paper between layers of wood in the sides of the frame house to help keep out the wind. If he is building in a very cold climate, he may put another form of wood, a building board, into the wall to help keep the house warm in winter. If he is building in a very warm climate, he may use the same sort of building board to help keep the house cool in summer.

These building boards may be made from the wood of trees or from some other plant material that can be used to serve the same purpose. Some are now made from the stalks of sugar cane after the sugar has been removed. Cornstalks, oat straw, and still other materials are also used in some kinds of building boards.

Since the purpose of building boards is to prevent heat from passing through the house walls, they are made of substances that do not *conduct* (carry or transmit) heat well.

If you place the bowl of a silver spoon in a cup of boiling water, the handle of the spoon soon becomes so hot that it will hurt your fingers to touch it. Silver and many other metals are good conductors of heat. If you put one end of a piece of cornstalk or straw or wood into boiling water, however, you can touch the other end with no discomfort whatever. Such substances do not transmit, or carry, heat well. So they are called *non-conductors* of heat. (*Non* means "not.") They are also called *heat insulators* because they keep heat that is on one side of them apart from or separate from places on the other side. (You will meet the word *insulator* again on page 163 of this book.)

The Mason Helps the Carpenter

If you watch the different craftsmen who construct a house, you will be interested to see how the work of one man fits into that of another. Even when a carpenter builds walls of wood, he needs a mason to make a foundation for him. He needs a mason, too, to build the chimney. The carpenter puts up the framework, the roof, the windows, and the doors and lays the floors. Then it is time for the plasterer to finish the walls, before the carpenter returns. At last he puts in the inside woodwork that is partly for show. He uses his plane and sandpaper. He smooths and polishes until some of the wood he leaves in sight may be very beautiful to look upon.

Wooden Ornaments

Wood, the material of the carpenter, no less than stone, one of the materials of the mason, is used, as you may have noticed, for ornamental as well as useful purposes. A long while ago people learned to fashion wood into attractive shapes. In the great tombs of Egypt, many pieces of carved wood have been found. Also in China, India, and Japan, there are to be seen specimens of wood that were carved with great skill long, long ago.

But no buildings a carpenter can erect and no ornaments an artist can carve are as beautiful as the trees themselves before they were cut to be used by men. So there is reason for people who appreciate the beauty of living trees to feel that some trees should not be cut.

Saving Some Trees

There are, indeed, many people who believe that there should be places where trees may live unharmed. You will be interested, in this connection, to read about national parks. You will like to know the reasons that led to their being set apart, by law, from land that is privately owned. You may learn that not all people have the same ideas about the treatment of trees even in our national parks.

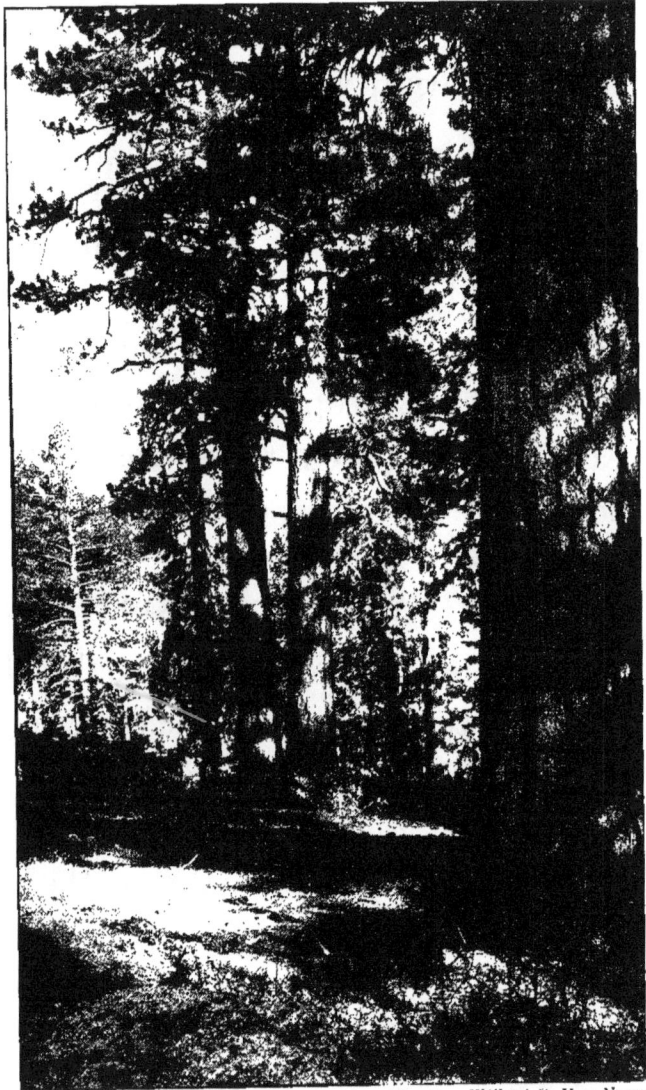

Man cannot make any wooden things so
beautiful as the trees themselves.

There is much to be considered about the proper and right use of forests by mankind. It is a matter for grave thought, also, how many magnificent trees should be saved that they may continue to make beautiful the places where they grow.

Many kinds of trees live to be hundreds of years old. Some kinds may grow for thousands of years. It takes but a short time to cut and use such trees; but if others ever grow so large to take their places — there will be a long, long wait before that time.

The carpenter can build a house
  With floors of birch or other wood,
And beams of oak and walls of pine —
  A shelter that is warm and good.

But carpenters cannot make trees
  Or floors such as the forests know
Or halls like pathways through the woods,
  For things like these must grow.

I like the shelter of a house,
But better, far, I love the trees
With trunks that stand against the wind
And leaves that whisper in the breeze.

(From *First Lessons in Nature Study*)

QUESTIONS AND ACTIVITIES

Lumber from the Pine Family

The word *conifer* means "cone bearing." Pines and firs and spruces and related trees that belong to the Pine Family have their seeds growing in cones. These trees are called conifers. Most, but not all, trees of the Pine Family have narrow leaves with straight edges. (An arbor vitæ is a conifer that does not have such leaves.) Most trees belonging to this family do not drop their old leaves in the fall. They wait until the new leaves have grown in the spring. So their boughs are always green. Such trees are called *evergreens*. (The larch, or tamarack, is a member of the Pine Family that is not an evergreen.)

Find, in a library, *Forest Facts for Schools* * or some other book about trees and read about trees of the Pine Family and look at their pictures.

* See page 424 for book list.

Photo by W. R. Mattoon, courtesy U. S. Forest Service

Shortleaf pine is an important timber tree.

You may like to read some of the pages in Part VIII in the book *Trees*. This part is about "The Cone-bearing Evergreens."

In the book *Our Plant Friends and Foes* there is a chapter called "The Pine Family." You may enjoy this.

Lumber from Broadleaf Trees

Maple, birch, oak, and many other trees are called *broadleaf* trees because their leaves are wide

Photo by W. R. Mattoon, courtesy U. S. Forest Service

A large white oak

and flat and different from the narrow leaves of pines and other conifers.

In the North, where there is much snow in winter, most trees with broad leaves drop their leaves in the fall of the year. In warmer parts of our country certain broadleaf trees, such as holly and live oak, do not drop their old leaves in the autumn. Such trees are evergreen.

Find a book about trees in a library and look up the following items:

(1) In what parts of the country does the white oak grow? Why is it called *white?*

(2) Make a list of ten broadleaf trees having wood that a carpenter is likely to use. Name at least one use to which a carpenter might put wood from each of these ten trees.

(3) Name five broadleaf trees that are not evergreen.

If you can get some pieces of different kinds of wood from a carpenter or a mill, make a chart by fastening the pieces to a sheet of cardboard. Print the name of each kind and notice the color of the wood.

WATCHING THE PAINTER

Protecting Metal Parts

After you have watched a mason lay the foundations for a house and a carpenter put up the wooden parts, you will look forward to the visits of a painter. Indeed, the painter does not always wait until the carpenter is through with his work before he makes his first visit.

There are likely to be pieces of iron used in places in the framework or the walls. The sooner this metal is covered with paint, the safer it will be; for it needs protection from moist air. Exposed iron, as you doubtless know, soon becomes coated with *rust*. Rust is formed when oxygen in the air combines with iron. Another name for rust is *iron oxide*.

Usually there are sheets of zinc or copper near the chimneys between the bricks and the

c, Charles Phelps Cushing

The painters cover the wood.

woodwork. Zinc does not need painting so much as iron does, but paint protects it so that it lasts longer than it would if left un-covered. One reason copper is so widely used in roofing work is that it does not need painting at all.

The first coat that the painter spreads on the iron work of a house is likely to be bright red. You may have noticed paint of this color on iron bridges and iron fences, too. Such paint is made with red lead and linseed oil (page 111). Quite often aluminum paint is used instead of red-lead paint. This is much the same color as silver.

Unpainted Wood

There are people who do not like to look at any kind of house paint. They say that the roof and sides of a house are more attractive if they are left to "weather" than if they are hidden under paint. They think clap-boards look best when they have turned as gray as the bark on trees. They consider a roof most beautiful after gray-green lichens or soft dark green mosses grow on the shingles.

What happens to such a house? Well, the side boards and the shingles keep changing in size. They swell and shrink.

In wet weather the wood absorbs moisture from damp air, or from rain or snow, and then it swells. At such times each piece of exposed wood pushes against the other pieces it touches and crowds against the heavy timbers to which it is nailed. It may even loosen the nails that were put in to hold it in place.

Then comes a dry spell. The hot sun shines on the wood and the moisture goes off in vapor. Every piece of exposed wood shrinks. It no longer presses hard against its neighbors but pulls away from them, leaving spaces between. It may rattle a bit in a strong wind and loosen the nails a little more.

During winter days wet snow may settle among the shingles. This may freeze at night and crack them.

So during dry days and wet days, through

Photo by A. G. Robinson, courtesy Charles Scribner's Sons

An unpainted house

Notice that the clapboards have *warped*, or twisted out of shape.

summer heat and winter cold, the unpainted house "weathers." Its mossy roof has a certain sort of beauty, but it leaks. The side boards are soft gray like bark, but they are loose and split and falling apart. The unprotected wood decays and crumbles.

Paints for Wood

You see, then, that anything which will prevent pieces of wood from changing size (swelling and shrinking) will help keep a wooden house in good condition.

Oil helps greatly to do this. Often shingles are dipped in oil that has been colored gray or green. The oil soaks into the pores (tiny openings) of the wood and helps preserve it. Sometimes people have the clapboards on the sides of their houses stained with oil, too. Wood soaked with oil lasts longer than untreated wood. Moisture cannot get into it so easily.

Paint, however, protects wood better than oil used alone. It is thicker than oil and forms a film over the surface. Water cannot easily get through the coating of paint unless there are cracks in it.

An excellent white paint is made with white lead and oil. The dark metal lead is changed to a substance of white color by treating it with an acid. This acid is the same sort that we have in vinegar. Most farmhouses in New England are protected with white-lead paint. It lasts well and does not need to be covered with a new coat so often as some other paints. There are, however, some good white paints that are made of other materials than white lead and oil. A white house looks clean and neat among the elms or other shade trees that are often planted near it.

Sometimes groups of people form societies of neighbors and agree to have no houses

near them that do not look well together.
They try to have all the houses near one
another in colors that are harmonious.   But

Courtesy Extension Service, University of Maine

A white farmhouse in Maine

in many places a man paints his house without considering the tastes of his neighbors
— he may even have a bright blue house next
door to a bright yellow one if he wishes to
do so.

Pigments

Any coloring material that is used in paint is called a *pigment*. White lead is called a white pigment because it makes a white paint. Zinc and certain other metals are also used as pigments.

Colored pigments may be added to white lead to give different colors. Some of the pigments are earths that have different natural colors. Red or brown or yellow earth may be dug and purified and then mixed in the paints. Indigo and some other plant colors have been used also.

During recent years chemists have made pigments in factories. They take certain products of black, sticky coal tar to work with. You may know that many beautiful colors and shades which the chemists make from these products are used in dyes. Some are also used in certain house paints, especially when bright colors are desired.

Linseed Oil

The first coat of paint that a painter puts on new wood is called a *priming* coat. It does not have much pigment in it. It has a great deal of oil. The priming coat fills the pores in the wood. Later the painter covers this with two or three coats of paint with more white or colored pigment and not quite so much oil as the priming coat had.

A good paint forms a film that is hard enough but not too hard. Most materials expand (become larger) when they are warm and contract (become smaller) when they are cold. The materials in the outer surface of a house expand a little when the hot summer sun shines on them. The same materials contract a little when the freezing winter air touches them. The paint film, if it is going to protect the wood, must not crack. So the film must be just right to expand without breaking when it is hot and to contract without breaking when it is cold.

The best known oil that permits paint to expand and contract without cracking or tearing the film is made from the seeds of the flax plant. We do not call it flax oil, however. We call it linseed oil. (The Latin name for flax is *linum* and one English name is *linseed*.)

Linseed oil forms its film by becoming hard. We say a paint is "drying" when the film is forming. But this process is not really drying. We call it drying because the paint becomes less sticky and finally hard. But wet things lose moisture when they dry, and the paint has lost no moisture. It has done something else. It has taken up some oxygen from the air. The oxygen helps the oil to form a new substance, which is the hard, glossy film most paints have after being exposed to air for a short time.

Driers and Thinners

If the painter wishes his paint to "dry" quickly, he uses chemical compounds called *driers*. When these are added to the paint, they cause the oil to take oxygen faster from the air. Thus the paint hardens more quickly.

A paint with just pigment and oil does not flow easily from the brush. So the paint is thinned with something. Turpentine is most often used for a *thinner*.

Turpentine is a substance that oozes from wounds in certain cone-bearing trees. American and French and Russian and Swedish turpentines come from different kinds of pine trees. Canada turpentine is taken from the balsam fir tree. This kind of turpentine is called *Canada balsam*.

As it comes from the tree, turpentine contains *resin* (rosin) and oil. People separate these two substances by a process called *dis-*

tilling. In this process the oil is changed to vapor by heating the turpentine in a closed vessel. The vapor rises and leaves the resin

Courtesy U. S. Forest Service

A turpentine orchard in Georgia

in the bottom of the vessel. The vapor is then allowed to pass through pipes and is cooled to liquid oil. After the turpentine oil is removed from the resin in this way, it is ready to be used as a thinner to make paint spread easily and smoothly.

Millions of gallons of turpentine have been taken in a single year from long-leaf pine trees in the southeastern part of the United States. These trees grow in sandy soil. They may be found in Virginia, south to Florida, and west as far as Texas. The long-leaf pine tree is important, too, for its timber.

Turpentine is also taken from young short-leaf pine trees. On page 99 you were told of another use for old trees of this kind.

The Painter Indoors

When the painter comes inside the house to work, he may use nearly the same kinds of

paint that he put on the outside. He may
paint the woodwork with paint containing
white lead, linseed oil, turpentine, and a drier.
For inside work, however, he uses more turpentine so that he may have a thinner coat.
And he uses more drier so that the coat will
harden quickly. He may also use any
colored pigment for indoor work.

Besides paint, there are other sorts of coats
for protecting and making beautiful indoor
woodwork. Some of these are not suitable
for the outside of the house, where they would
be exposed to all sorts of weather conditions.
So, if you watch a painter while he is inside
the house, you may learn about such different
materials as varnish and shellac and lacquer
and wax.

Varnish

When painters wish to have a shiny, hard
coat on wood, they sometimes use varnish.
A coat of this sort is transparent — that is,

you can look through it and see the surface of the wood.   It does not hide the wood, as paint does.

Varnish is made by dissolving certain gums or resins in some substance.   There are three sorts of varnishes, depending on what is used to dissolve the gum or resin.

One kind is called *spirit varnish*.   This is made by dissolving the gum or resin in alcohol.   (Alcohol and liquids with alcohol in them are often called "spirits.")

A second kind is called *turpentine varnish* because it is made by dissolving gum or resin in oil of turpentine.

Oil varnish is the third kind.   Some oil that hardens with a good firm film is used — such as linseed oil, poppy-seed oil, or tung oil.   These oils may be used alone or they may be mixed with resin and turpentine.

Tung oil is made from the seeds (nuts) of tung trees.   For a long while we bought

all our tung oil from other countries where
these trees grow naturally.   But thousands
of acres in southern states of our country

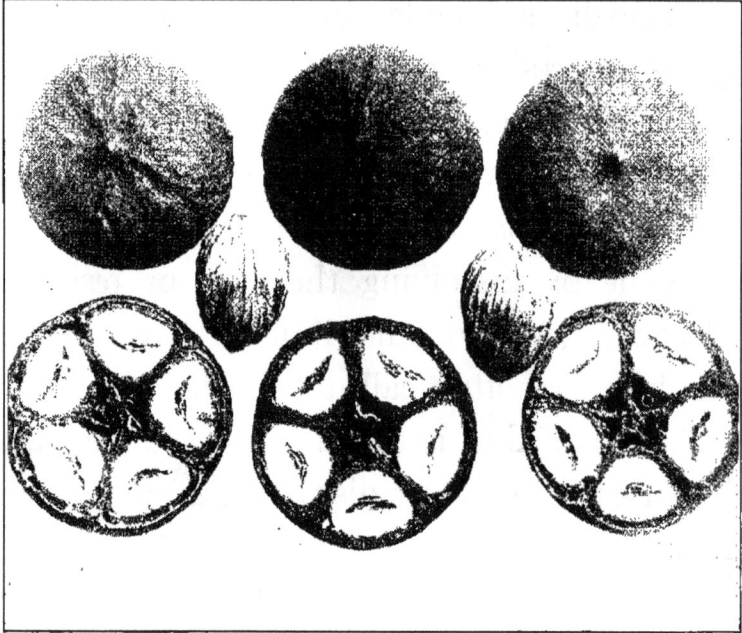

Courtesy U. S. Dept. of Commerce

Tung fruits

Three are cut to show the seeds inside.

have been set with tung trees.   So now tung
oil is made in the United States, too.
Another name for the tung tree is *varnish
tree*. It belongs to the Spurge Family.

You will now read about some varnishes that are known by the special names of shellac and lacquer.

Shellac and Lacquer

The *lac insects* belong to the family called Scale Insects. Many kinds of small scale insects have wax glands in their bodies with pores through which the wax oozes. This wax hardens like a little shell or scale on the outside of the body of such an insect.

Lac insects live in the warmest parts of Asia and some other places. They suck the juices of fig trees and some other trees. They produce so much wax that it is gathered and sold. This is called lac.

Lac will dissolve in alcohol, and this solution is called shellac. The painter has a number of uses for this substance. It is very good for coating knots in new wood. This prevents the gum (pitch) in the knot from working through and staining paint when it is put on later.

One kind of lacquer is also made with lac. This is used to cover brass and iron and some other metals. It keeps them from becoming discolored or rusty.

But a kind of lacquer that has long been famous as a durable hard varnish for wood is made with the juice of the lacquer tree. The lacquer tree is related to sumac bushes. You may have seen some beautiful boxes from Japan covered with lacquer. This sort of varnish is used with different colors.

The lacquer you are most likely to see a painter use, however, is a third kind; for chemists have learned how to make a new sort of lacquer. They take the short fibers from cotton seeds and prepare them so that they can dissolve in certain chemicals. This sort of lacquer can also be colored and used as a hard varnish. It is good for covering metals and some other substances. You may see this kind of lacquer on automobiles.

Wax

Often floors and other woodwork are not varnished. They are sometimes polished with wax, instead. If the wood has a beautiful grain, it may look best when treated this way.

Beeswax is a common animal wax, and the most common mineral wax is called *paraffin*. There are many uses for both these waxes, and they are often used for polishing wood. For this purpose they are kept rather soft so that it is easy to rub them over the wood.

Natural Uses of Painters' Materials

You know that certain birds and insects are natural masons. You know that some birds and insects work with wood, which is the carpenter's material.

Did you ever notice any of the ways that many plants and animals have of using some of the same sorts of materials that a painter uses? Only a few of these ways can be mentioned here. You may learn of others for yourself if you care to do so.

As you know, the sap of pine trees (and other trees of the same family) is a mixture of resin and oil. One use these trees have for this turpentine (or pitch) is to dress their wounds with it. If the bark is broken so that the wood inside is exposed, this sticky sap oozes over the wound. It hardens and forms a coating which protects this part from certain insects and from disease germs that might otherwise enter. It protects this part from weather injuries also.

Plants have other uses for wax. Many coat their fruit with it. You may have noticed a white coating on blueberries, grapes, or plums. It is easy to rub this off with your finger.   It is a dainty covering of fine wax particles.   Rain runs off the surface of these fruits.   It cannot soak in.   The wax is a protection against sun and drying winds, too.

If you take a cloth and rub an apple, it will be as shiny as a piece of furniture polished with wax.   This is because the apple actually has a coat of wax over its skin. Some kinds of apples have more wax than others.   One year an apple buyer in Chicago refused to take a carload of beautiful red apples that had been picked for him in western orchards.   He said the apples had been dipped in paraffin.   No one had dipped those apples, however.   The wax on their red cheeks was their own natural kind.

Bayberry bushes, as you know, put so much wax on their fruit that it is gathered and used for candles called bayberry dips.

Oranges, lemons, and similar fruits have oil in their rinds that gives them waterproof coats.

Many animals have some sort of oily waterproofing. Some oil is present on the hair of most mammals. It is secreted, or produced, by the skin. The hair of sheep is covered thickly with fatty oil. It helps prevent the hair from tangling. It is known as *wool wax*, or *lanolin*. After sheep have been sheared, this substance is removed from the wool. Lanolin makes a very good skin ointment.

The feathers of a bird are also covered with oil. This oil, however, is not secreted from the skin all over its body, as is the case with a hairy animal. A bird has the task of keeping its feathers oiled. It has an oil gland on its back near the base of its tail. This gland secretes the oil. When a bird is ready to dress its feathers, it presses out a drop of oil from the gland with its beak. Then it passes its beak over its feathers, one by one, leaving a little oil on each. We say a bird is *preening* when it does this.

Most land birds have small oil glands. All swimming birds have rather large oil glands, as they need more oil. You can see why water runs off a duck's back so readily.

You have learned that the lac insect and some other scale insects secrete wax which forms over their bodies in shell-like covers.

Most aphids secrete wax from wax glands, also. Aphid wax is usually white and powdery. It does not fit over the body like a shell. Often it looks much like the waxy "bloom" on blueberries or grapes.

Some species of aphids, however, secrete so much white, fluffy wax that they have a covering that looks like wool. These are sometimes called *woolly aphids*. They live in colonies on certain plants. One species is quite common on alder. Another is often found on apple trees. The colony in the

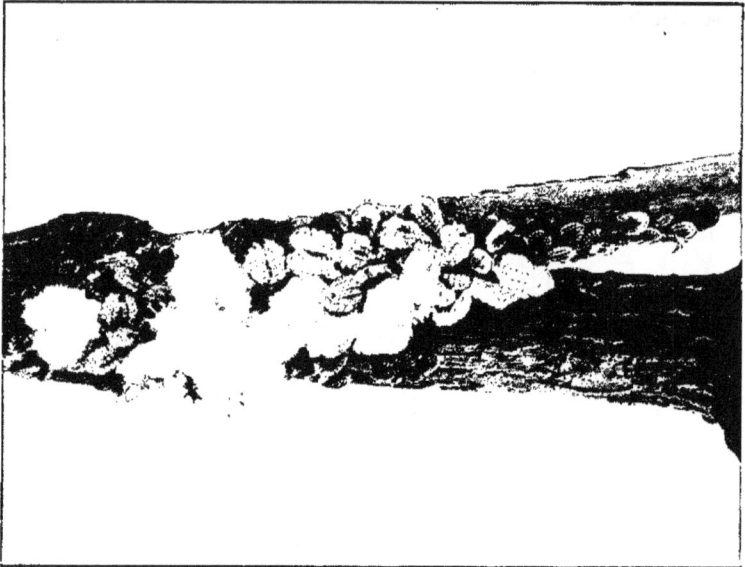

Courtesy Maine Agricultural Experiment Station

A colony of woolly aphids

picture is a third kind, which lives on ash trees.

Many caterpillars spin cocoons in which to rest while they are in the pupa stage. After a caterpillar finishes spinning, it usually washes the inside of the chamber with a liquid that hardens and becomes rain-proof.

So you see that, when the painter puts a protecting coat over wood, he is using materials similar to those used by many plants and animals.

QUESTIONS AND ACTIVITIES

Look on the pages in this chapter for all the words that are printed in *italics* (letters that slope as in that last word). Copy all these words on paper. Then write as many sentences as you have words or expressions, using one of them in each sentence. If you do not remember what one of them means, read about it again.

On page 118 you read that the tung tree belongs to the Spurge Family. Did you ever read about any other useful tree that belongs to this same family of plants? In this connection ask in a library for the book *Through Four Seasons* * and read Chapter Seventeen, which is called "Tapping a Rubber Tree."

*See page 424 for book list.

An Exhibit

If your teacher is willing, have an exhibit at school. You might include samples of the following :

(1) A piece of wood coated with clear varnish

(2) A piece of wood coated with varnish stain (varnish that has been colored so that it can be used to color or "stain" wood)

(3) A piece of hard wood polished with wax

(4) A piece of unpainted wood that has been exposed to the weather for more than a year

(5) A piece of painted wood that has been exposed to the weather for more than a year — if you can find any that can be spared

You might also have samples of the following :

(1) Fruit with a waxy coat

(2) Fruit with an oily rind

(3) Some wax made by insects

Beeswax

On page 121 beeswax was mentioned. This is the most important kind of animal wax that people use. Read in some other book about this sort of wax. Ask in a library for the book *The Bee*

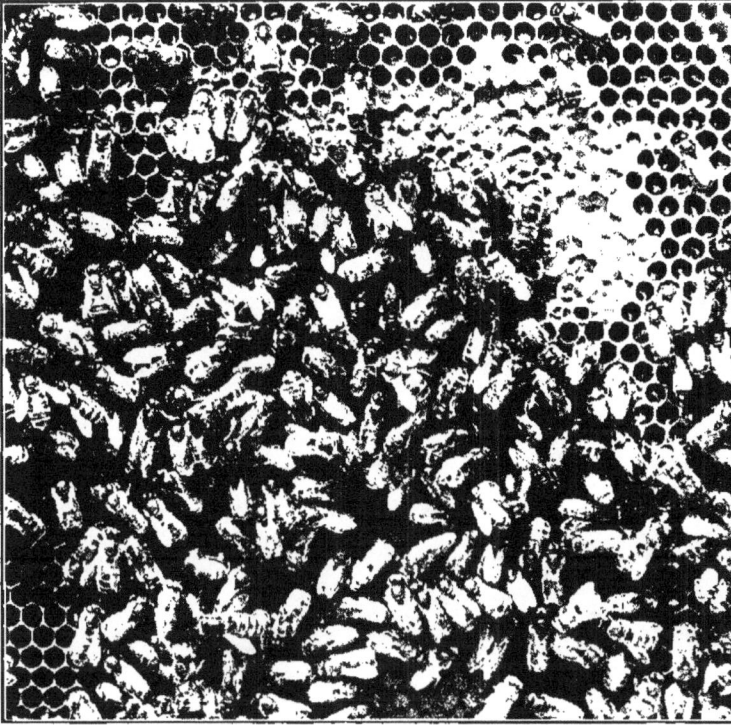

Courtesy U. S. Dept. of Agriculture

Honeybees making a comb of beeswax

People. In that book read Chapter XIII, "The Work in the Hive — The Manufacture of Wax." Read also Chapter XIV, "Honey-comb."

On page 127 you were referred to *Through Four Seasons.* Look at the picture on page 113 of that book and read on page 112 about the sticky resin with which the overwintering buds of trees and shrubs are varnished. This substance is called *propolis*, or "bee glue." Honeybee workers gather propolis and use it to varnish the inside of the hive. Write a short essay with the title "Resin Used by Trees, Bees, and People."

WATER·LIGHT·HEAT

PART
THREE

Water tower at Towson, Maryland

THE PLUMBER

The word *plumb* comes from a Latin word that means *lead*. At first the name *plumber* was given to a man who was a *worker in lead*. His duties were to cover the roofs of important buildings with sheets of lead and to keep that lead in good repair. He had little, if any, work to do inside a house. That was a long time ago.

But, if you hear your father say, "It is time for the plumber to come to put the plumbing into the new house," you know that he does not mean it is time to put on a roof. He is thinking about the pipes through which water may be brought into the house and about the pipes through which waste, or *sewage*, may be drained from the house. He may be thinking, too, of pipes

133

for hot water or steam. All these modern matters are sorts of work with which the first plumbers had nothing to do.

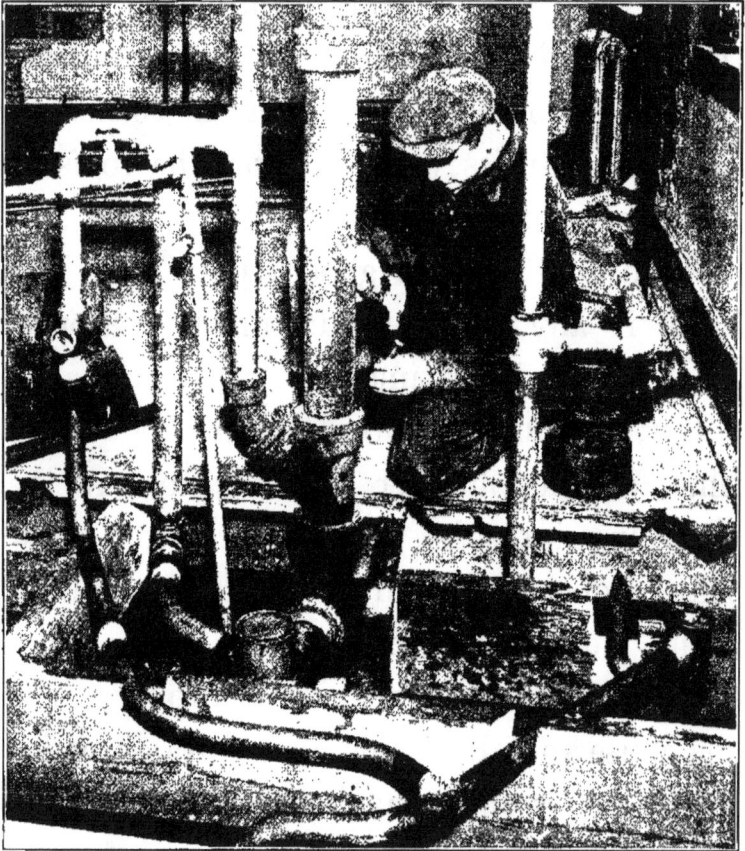

Courtesy "Lead," March, 1933

A plumber at work

The oldest kinds of metal pipes were made of lead. Although many pipes are now made of iron, copper, and brass, lead pipes are still used for some purposes. So a modern plumber is true to his old name — for he can still work with lead. Even brass pipes may need to have their joints held together with melted solder (lead mixed with tin). Solder holds metals together because, when it is heated, it unites with their clean surfaces. Such a mixture of metals is called an *alloy*.

Water for Country Houses

A country house must have its own water system. The water is usually taken from a well. It would often be possible to get water enough by digging a hole in a low place where little streams could run into it from the surface of the ground. Such surface water, however, is not safe to drink. It may become unclean in several unpleasant ways. It is quite likely to contain some dangerous disease germs, such as those causing *typhoid fever*, which are often carried by surface water.

It is necessary, therefore, to dig a well in a place where underground water may flow into it. You know that brooks run down the sides of hills into larger streams or rivers that flow into lakes and the sea. There are also streams that move under the ground out of sight. Sometimes a stream of this sort may break through the ground on a hillside or other slope and form what we call a *spring*. Usually, however, we do not see water from an underground stream unless some one digs a hole deep enough to reach it.

In making a well, a man digs through several layers of earth. On top there is the *soil* layer, in which plants grow. Surface water from rain or melted snow can easily soak through this top layer. This is called a *porous* layer because it has pores through which water may pass.

Beneath the soil layer there is often a layer of *clay*, which may be very thick. Water does not seep (soak) through heavy clay at all readily. It is called a *nonporous* layer. The clay layer may have been folded and broken, however, by an earthquake or through some other cause. In that case water could seep through the broken place and reach the layer beneath the clay.

Perhaps the third layer will be *sand* and *gravel*, through which water can seep since such a layer is porous. Under the gravel there may be a nonporous layer of rock called *bed rock*.

There are high places and low places in the bed rock, just as there are hills and valleys on the surface of the earth. The underground streams flow down the slopes of the bed rock as surface brooks flow down a hill-side.

The water of a deep underground stream is usually safe and wholesome to drink. Although it was once surface water, it has had

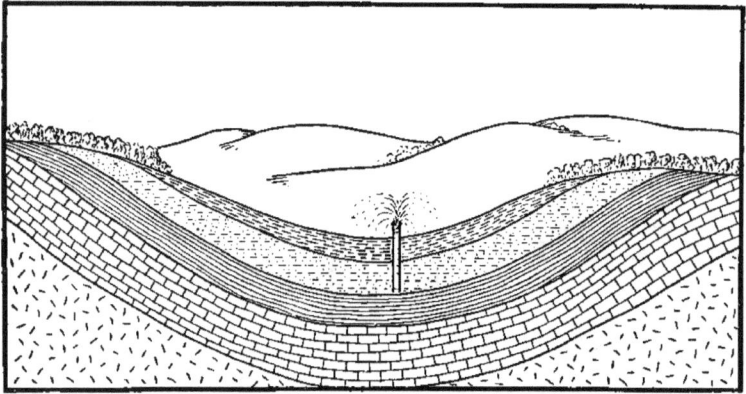

What causes the water to gush like a geyser?

to seep, or *filter*, through porous layers. Impurities in water are removed in this filtering process.

When a man digs or drills a well, he may reach an underground stream that flows between two layers of nonporous clay or rock. The water may be coming from a much higher hill-like slope of bed rock many miles away. Then, as soon as an opening is dug or drilled into the channel of the stream, the water will gush into the air like a geyser or a fountain. The pressure of the higher distant water forces it up. Such a well is called an *artesian* well.

Most wells are not of this kind. The water does not come up like a fountain but remains in a quiet pool in the bottom of the well. It must be drawn up in some way. This may be done by letting down a bucket tied to a rope and then pulling up the bucket full of water. It is slow and hard work to do this by hand. People learned long ago how to pull up the bucket more easily by using a *well sweep* (also called a *balance pole*).

If a pole, with a rope (or chain or stick) and bucket fastened to one end, is so placed that it is almost balanced, the full water bucket may be raised with very little effort.

A well sweep, or balance pole, is a kind of *lever*. In the chapters on tools, in Part Four of this book, you will read more about levers and how work is done with them.

This old house has a well with a well sweep.

Although some wells with well sweeps are still in use, most wells today have suction pumps. A pipe is placed with its lower end in the water in the well, and a pump is attached to the upper end. A suction pump, which is also called an air-lift pump, depends

upon the pressure (weight) of air to bring
the water up from the well. There is a disk

A suction pump is worked by hand.

(flat circular piece) of leather or other suitable material which fits very tightly inside

the pipe. This disk is fastened to a rod that is moved up and down by the handle of the pump. Some of the air is sucked out of the pipe by the disk when the rod moves up. This causes a partial *vacuum* in the pipe.

(The word *vacuum* comes from the Latin word *vacuus*, which means *empty*. A vacuum is a space from which the air has been removed. A partial vacuum is a space from which part of the air has been removed.)

There are two *valves* (parts that open and shut like little doors) in the pump. These open to let the air out of the pipe and close to prevent other air from getting in.

You see, there is air pressing down on all the water in the well outside the pipe. The weight of this air forces water to rise inside the pipe, from which the air is being removed. Every time the pump handle moves the rod down and up, more air is sucked out by the disk. So the water rises higher and higher in the pipe until it passes the valves and at last comes out the spout.

All this time that we have been talking about wells and outside pumps we have said nothing about any work by a plumber inside the house. A suction pump, however, may be placed in the kitchen and have a long pipe reaching to the water in the well. Such a house may be said to have the very simplest water system and plumbing — a single water pipe. This arrangement saves carrying the water into the house from an outside pump. It is pleasanter to use in cold or stormy weather.

It is rather hard, slow work to pump, by hand, all the water that is needed even when the pump is in the house. For this reason many country people have gasoline or electric engines to do the pumping.

In a better private system of waterworks there is a large water tank which may be filled by a pump (worked by hand or by an engine). This tank is fitted with pipes leading to the bathroom and to faucets in the kitchen.

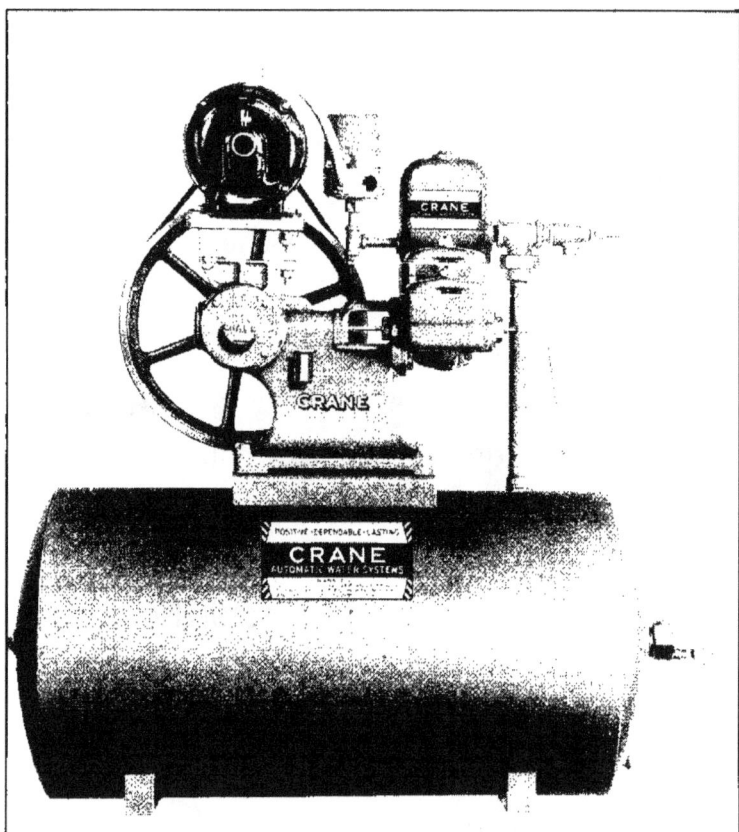

Compressed-air tank, pump, and electric motor

A water tank may be placed in the attic, and then the water runs "downhill" through the pipes. Or, if it is a *compressed-air tank*, it may be placed anywhere — even in the cellar.

A compressed-air tank is a strong metal air-tight tank. Water is pumped into it until the air is pressed together in the top of the tank. Air is elastic. It can be forced into a small place; but when it is released it quickly expands, pushing its way out. When water is pumped into a closed tank, the air in the tank is pressed into a smaller and smaller space. Then, if a faucet is opened, the water in the tank is forced out by the expanding air.

Instead of tanks of these sorts, there may be tanks to hold rain water. They are called *cisterns*. In cold climates, cisterns are placed underground or in the cellar. They are usually made of concrete and are often connected with pumps.

The water that falls on the roof of the house is taken through pipes to the cistern. But before it is allowed to pass into the cistern, it must run through a layer of clean sand or charcoal to take out any dirt that may get into the water from the roof or the pipes.

In warm parts of the country, wooden tanks built above ground may be used for cisterns. If such tanks were left open, insects would get into them. They would be breeding places for mosquitoes. In Louisiana and some other places, people have been forbidden by law to have uncovered cisterns. They must at least be well screened.

You can see that a country house with plenty of good pure well water, a water tank, and an engine to do the pumping may be as well equipped for its water supply as any city house. But, of course, this system of waterworks is a private one. The owner of the house has all the expense of putting in such a system and keeping it in good condition. He does not pay a city or company to furnish him water, but his water costs him something for all that.

Wastes from Country Houses

Many country houses have no pipes for carrying away waste water. So all the water from the dish pan and washtubs and other wastes must be taken from the house in pails. Some of it is buried and some thrown on top of the ground.

A country place, however, that has plenty of water piped into the house usually has a pipe to carry away waste from the kitchen sink. It may also have a bathroom with all the convenient pipes.

But no town sewer pipes run by the farmer's house to take away such wastes, or sewage. What, then, becomes of it? A country system of taking care of sewage, like its waterworks, must be a private one. The owner must find ways to dispose of the sewage. Sometimes he has round tile pipes laid to a brook. Sometimes he digs a great hole, called a cesspool, to receive the sewage.

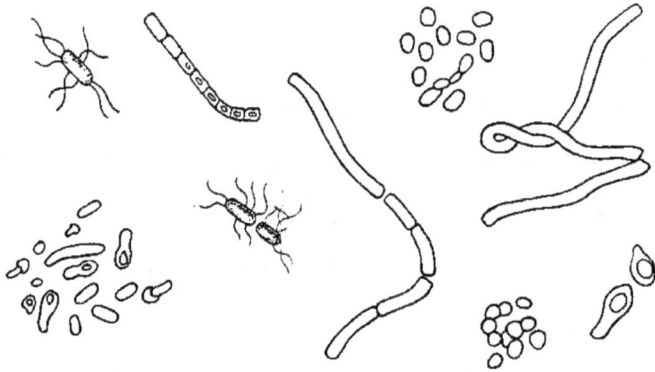

Some cesspool bacteria, much enlarged

The cesspool may be kept safe for use by adding certain chemicals or by allowing some kinds of *bacteria* to work upon the sewage.

A *bacterium*, as you may have learned from other books, is an extremely small form of life with only one cell (simple part) in its body. There are many kinds of bacteria that are harmful to the health of plants and animals. Certain kinds, however, are helpful in various ways to other forms of life. The work that some bacteria do in a cesspool is to use the filthy matter there for food and thus get it out of the way.

There are other methods of treating wastes. Some of these methods, which different cities use, are mentioned on pages 153–154. These same ways are sometimes used in disposing of the sewage from country houses.

City Waterworks

A city must have, on a very large public scale, what the best-equipped country house has on a small private scale — pipes to bring fresh water in and pipes to carry away the sewage.

City waterworks may be owned by the city itself or by some service company. All

the house owners need to do to obtain water
is to have their own pipes connected with the

Water tower for the city of Indianapolis

larger pipes which carry the city water.   And
of course the people who use this water must
pay a certain amount of money for it.

Some towns and cities get their water from powerful springs or deep wells, but most city water comes from rivers or lakes. People try to get the cleanest water that is near the city. Sometimes it is carried many miles in pipes.

Unless it is considered quite pure, water is filtered through layers of charcoal or clean sand. And very often it is treated with chemicals to make it safe to drink.

It is common to have a water *reservoir* on top of a hill if there is a hill in or near the city. The reservoir may be an upright cylindrical tank, called a *standpipe*. It may be in the form of a tower. Or it may be a huge concrete basin. When the basin is filled with water, it looks like a pond. In some of the pondlike reservoirs there are fountains which throw the water into the air in a fine spray. The air helps purify the water.

The air improves the odor and taste of the water by changing or carrying away some of the unpleasant results of decaying plant substances which are sometimes present.   Some

Courtesy Malcolm Pirnie, Consulting Hydraulic Engineer, N. Y. C.

City water being filtered and aired for Providence, Rhode Island

water has a form of iron in it that darkens its color. Oxygen from the air combines with this iron so that it settles, leaving the water clear and with a much better color. The air also drives out some of the gas called carbon dioxide, which plays a part in causing pipes to rust.

A reservoir holds a large supply of water from which the people get what they need each day. The water it holds is ready, too, to use as a help in fighting fires.

City Sewage

The waste water from city houses is taken care of by the city or by a service company. It goes from each building into large sewer pipes. There are a number of different ways of disposing of sewage, a few of which will be mentioned here.

(1) Some cities let their sewer pipes empty into streams, lakes, or the sea.

(2) The sewage of some cities is strained through layers of charcoal or ashes. The liquid parts then drain away.

(3) Certain chemicals are put into the sewage in some places. These help purify it. *Lime* is one of the substances used.

(4) In some cities helpful bacteria are used. The sewage is drained through soil where these bacteria are abundant. The bacteria destroy the materials that give rise to the bad odors of the sewage and help purify it.

(5) Recently some cities have built large numbers of concrete vats or tanks in which conditions are made favorable for the work of helpful bacteria. When the bacteria have completed their task, the materials that remain are filtered through sand or in a machine. The water is then safe to be emptied into lakes or streams. The solid material is sometimes dried and used as a fertilizer.

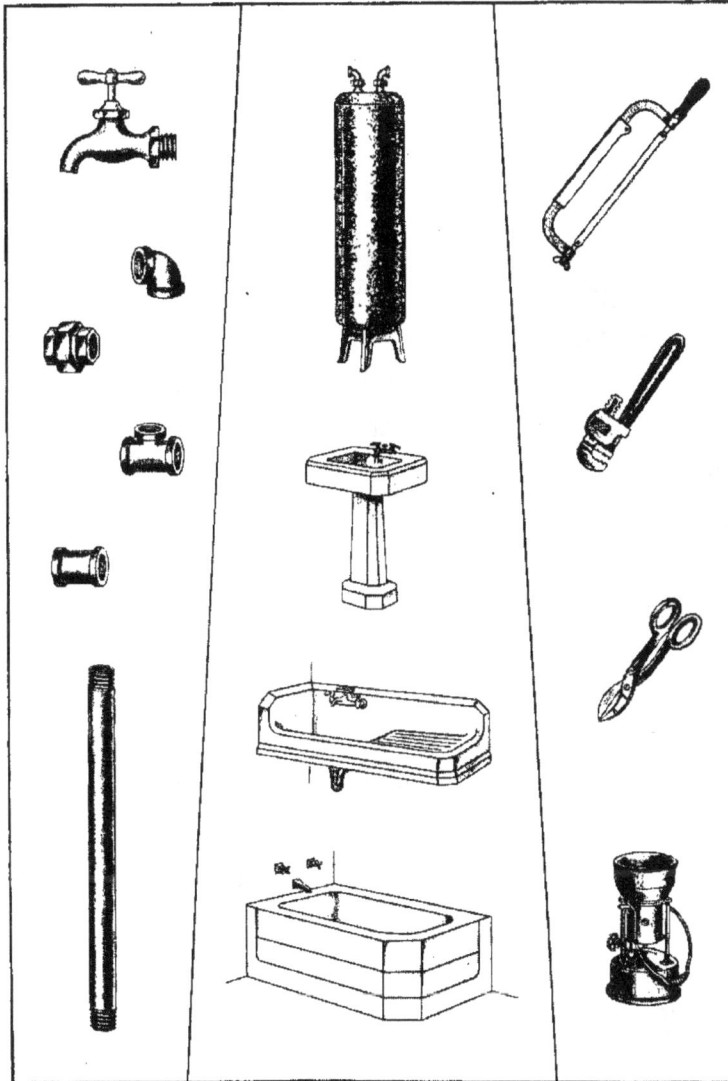

What does a plumber have to do with each of these?

Of course people living in the country can use these ways, too.

QUESTIONS AND ACTIVITIES

On page 155 are pictures of things that plumbers put into houses.　Look, in your home or at school, for as many of these things as you can find.

On that page, too, are pictures of some tools that plumbers use.　Do you know how he uses them?　If you do not know, ask some one to tell you.　Perhaps your father or a plumber or a man in a hardware store will tell you.

Where does a plumber put small rubber or metal circles called *washers?*　Can you find one of these in place at home or in school?

You read a little about bacteria in this chapter. Look in the index of the book *Through Four Seasons* * for the word *bacteria*.　Read all you can find about them in that book.

Give a short talk about "A Plumber and His Work," or write a little about the same subject.

*See page 424 for book list.

THE ELECTRICIAN COMES

Electric Lights and Flame Lights

Most of you boys and girls who read this book have never used the light of an oil lamp. Yet, when your grandfathers were your age, light from oil lamps was the commonest kind.

Such lamps needed considerable care to keep them in good condition. Your grandmothers, or other people in the homes, filled the lamps with kerosene (coal oil) usually as often as every day. They rubbed or cut the charred tips of the wicks so that the trimmed wicks could burn well and with an even flame. They washed the chimneys to remove any soot that had collected on the inside the night before. They polished the chimneys carefully with clean cloths so that the glass would be shiny and clear.

How would you like to go to all that bother before you could have a light to use? In

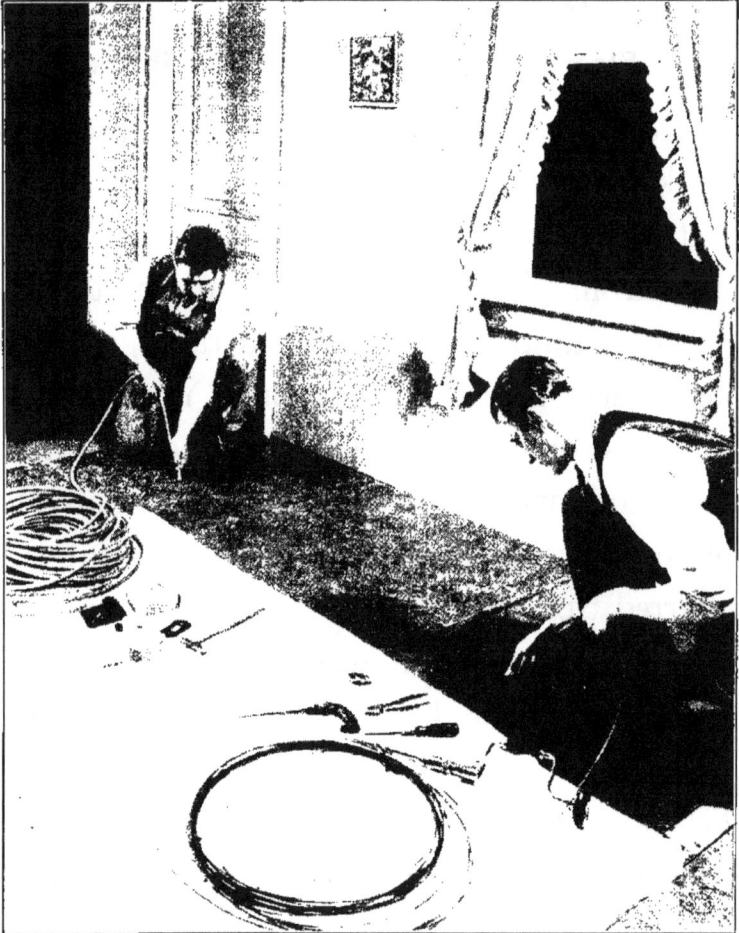

Placing electric wires

those days people were thankful enough to have such lamps. They would not have been willing to give them up and go back to the candles their fathers and grandfathers and great-grandfathers used.

We still like to see candles at special times and in certain places. But the flickering flames tire our eyes, and we should not care to do much reading by candlelight. We would rather have the tedious and smelly task of caring for kerosene lamps for the sake of their steadier and stronger light.

If you live in a house that has electric lights, however, you escape all that sort of trouble. You do not even need to strike a match when you wish a light. You simply push a button or pull a cord — and a dark room is suddenly lighted!

You get this sudden light because by pushing the button or pulling the cord you permit electricity to move along the fine wire, or *filament*, which is inside the electric bulb. The electricity causes the filament to become white hot and thus supplies the light.

An *electric current* (electricity in motion, like a stream) is constantly passing along the main wires with which a house is equipped. We call the path along which an electric current flows its *circuit*. There is a device called a *switch* which opens and closes that part of the circuit that leads to the electric light. When the circuit is closed, the electric current flows along the wire

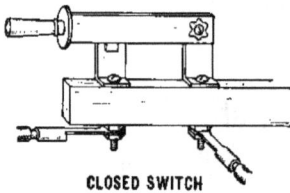

CLOSED SWITCH

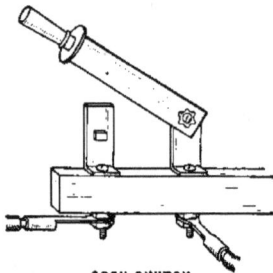

OPEN SWITCH

A closed and an open switch

connected with the filament and along the filament itself. When the circuit is opened, the electric current cannot pass the switch. By pressing the push button or pulling the cord, you operate the switch and thus close the circuit (when you turn on the light) or open the circuit (when you turn off the light).

There is not so much danger of fires when a house is properly wired for electric lights as there is when a house is lighted by flames (as by candles, oil lamps, or gas that runs through pipes).

Electric Shocks

Electricity, however, has dangers of its own. You should know what some of these dangers are. You should know better than to do what the person in the picture on page 162 is doing. He is turning on an electric light with one hand and turning on water at the faucet with the other.

You should never touch the metal around an electric light with one hand while you are touching "grounded" metal with the other. (We say the faucet is *grounded* when it is connected with a metal pipe that runs into the ground.) If you do this, a current of electricity is likely to pass through your body. You will then have an *electric shock.* **If** the current is not strong, the shock may only surprise you and make you jump. But if it is strong enough, it will kill you.

There is a chance for such an accident in some bathrooms and some kitchens and some other places where there are faucets or

Courtesy U. S. Bureau of Standards

Danger!

Do not turn on an electric light with one hand while you are touching a water faucet with the other.

"grounded" metal for other uses. You will wish to remember this.

Insulation

The electrician knows how electricity behaves and how to control it and use it. He knows how to keep it on the wires over which it travels. To keep it there, he must *insulate* the wires from other things through which electricity can run more readily. (The word *insula* means an *island*. To insulate a thing means to make it like an island — that is, to separate it or keep it apart from certain other things.)

Rubber is an excellent insulator. So, when the electrician comes to wire the house, he brings long wires that are covered with rubber. Sometimes there is a paper covering over the rubber. Sometimes silk or cotton or rayon fibers may be woven about the rubber to protect it and to improve its appearance. Notice also what is said about "conduits" on page 167.

Switches and Fuses

Plumbers use valves and faucets to control gas and water. The electrician uses switches and fuses (see page 165) to control electricity.

Rubber or porcelain, or some other substance through which electricity cannot easily go, is used with copper in the making of a switch. As we have stated, the current of electricity can pass along the wire while the circuit is closed; but, when the circuit is open, the current is stopped. It cannot then travel past the switch.

Electricity will follow through some metals much more readily than through others. It goes very easily along copper. For this reason copper wires are used so that the electricity may have a good path to follow.

When electricity is flowing along a wire or a bar, it may meet an object which resists it (which is very hard for it to pass). Then the energy that the electricity uses in trying to pass causes this object to become warm or even hot. You may compare this with the way your own body becomes warmer when you use extra energy in a bit of hard work or other exercise.

People make the most of this feature of electrical energy. By choosing a kind of wire that is not damaged by being heated and cooled many times, they can make toasters or electric heaters for different purposes.

This feature of electricity is useful in another way, too. It makes it possible to use the protective device called a *fuse*. A fuse is a piece of special metal which becomes so hot that it melts when the current of electricity flows at too great a rate. (By *rate* of current we mean the quantity of electricity that flows by a point in a given time — as so much a second.) During the time of spring floods, the current of water in a river is sometimes great enough to wreck

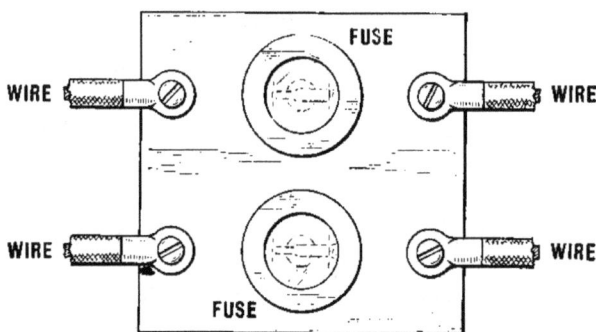

Wires with fuses

bridges. Similarly, in some cases the current of electricity may be great enough to wreck electrical apparatus. A fuse, however, prevents such a wreck. By melting, it breaks the circuit. Then no current at all may pass. That pathway is spoiled until a new fuse can be put in to take the place of the melted one.

Placing the Wires

The work of an electrician is not simple. It is somewhat like solving a puzzle to figure out the best way to run the wires through a house.

They must not be left where the insulation will get wet or be rubbed off; for, if the insulation is destroyed, the electricity may leak out through the break. Then, after a time, that part of the wire may become hot, like an overheated fuse. If it chances to rest against a piece of wood, this heat may cause a fire.

Sometimes wires are run through places in the walls or under floors where rats may find them. In such cases the rubber-covered wires are pushed through metal tubes which the rats cannot damage. These metal tubes, which are easily bent with suitable tools, are called *conduits*. Flexible, or bendable, *cables* are also made with insulated electric wires inside and metal outside. A cable differs from a conduit, however, in having the protective coating so firmly attached to the insulated wires that the wires cannot be pushed into and out of the coating, as can the wires inside a conduit. Cables are often twisted and look like metal ropes. Conduits and cables are used to protect electric wires in various places besides those where rats may enter.

You can understand how important it is that an electrician should do his work well. Faulty wiring means danger of shocks to people living in the house and danger of fires. Companies which hire electricians to do wiring are careful to employ only intelligent and dependable men for this work.

There are rules that regulate the work of electricians. One rule is that "each lighting circuit shall contain one wire which is grounded and connected to each lamp on the circuit." A common way to ground wires

Some things the electric servant can "run"

is to attach one end of the main wire to a
water pipe. The pipe goes so far into the
ground that this is a good way. The place
where the wire is fastened to the pipe is usually
underground in the cellar or basement.

The Electric Servant

When the electrician comes, he brings us a
marvelous servant. As you know, we push
a button or pull a cord to call this servant,
and it lights the room for us. (No lamps
to fill! No chimneys to wash! No wicks
to trim! Not even a match to strike!)

But this is not all. This same servant
will heat things for us — a toaster, a flat-
iron, a kettle or other dish, a cookstove, or
radiators to heat the rooms.

If we give this servant the proper sort of
refrigerator, it will help keep our food cold
and help freeze our ice cream.

Then, too, it will do many sorts of heavy

work about the house if we provide the necessary equipment. It can wash our clothes and iron them. It can also run a carpet sweeper to clean our carpets or rugs, and it

An old-fashioned carpet beater

can do a better job than a woman can do when she beats a rug with a carpet beater.

Electricity in Animals

In other chapters of this book you read about birds and insects that may be called masons and about some that may be called carpenters. Perhaps now you will like to read about animals that use electricity.

There is electricity in your own body. When you move your arm, an electric change

Courtesy New York Aquarium

Electric eels

takes place in the muscles of your arm. Electric changes occur in other muscles when you move them.

All animals, indeed, have electricity in their bodies. Most animals cannot make

any special use of this electricity.   They cannot give shocks to protect themselves.

There are some fishes, however, that have
really powerful electric organs.   They can

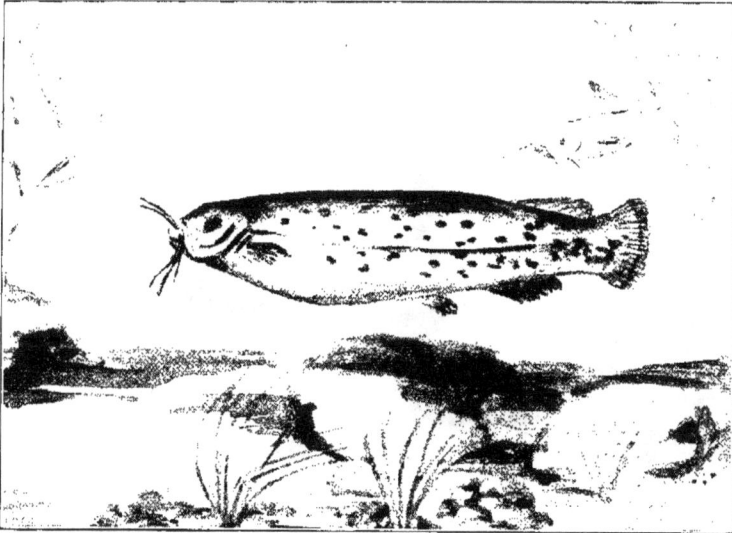

Electric catfish

protect themselves with these organs.   When
fishes or other animals try to catch such a
fish, it can give them electric shocks which
stun them.   All it needs to do is to touch
its enemies.   Then it can escape.   When one

of these electric fishes is hungry, it goes up to its prey and gives it a shock that stuns or kills it. Then it can eat its dinner easily.

Electric ray

The most powerful of electric fishes is called the *electric eel*. This may grow to be more than six feet long. It can stun animals as large as men, or larger. It lives in shallow water in the Amazon and some other South American rivers.

The *electric catfish* lives in the Lower Nile and some other rivers in Africa.

Electric rays (also called *torpedoes*) live in all warm seas.

Besides these fishes which we have mentioned, there are many others with electric organs.

QUESTIONS AND ACTIVITIES

Have you noticed that your eyes are not comfortable when you are in some rooms that have powerful electric lights placed where you look directly at them? Have you noticed that, in other rooms, you can see well enough and the electric lights seem pleasant? Your eyes do not ache in such rooms.

Look at different kinds of electric lights. Do some have shades that keep the glare from your eyes? Do some have reflectors under them that throw the light against a white ceiling? Describe an electric light that hurts your eyes. Describe one that does not.

Look in books for pictures of two kinds of electric lights in the sky — *lightning* and the *aurora*.

There is a picture of lightning on page 25 of *Holiday Hill*.* There is a picture of the aurora on page 186 of *Introduction to World Geography*.

Would you like to use this light if it had no shade?

In North America, the aurora is often called *northern light*. Do you know why? Where would you travel to see the aurora that is spoken of as *southern light*?

* See page 424 for book list.

Look on pages 285–298 in the book *Surprises* for some games with electricity.   If you have never played these games, you may like to do so now.

If the book *Working with Electricity* is in your library, read at least one chapter of it.

If you had an electric "servant," what would you like best to have it do for you?   You have read about some kinds of work that can be done by things that are run by electricity.   Try to name some other kinds.

Some words in the chapter you have just read are printed in *italics*.   Copy these words on a piece of paper.   Then write as many sentences as you have words or expressions, using one of them in each sentence.

Look at pictures of electric lights in papers, magazines, or catalogs.   Find some pictures of lights that you think would be good for a large public building.   Find some that you think would give a pleasant light in a room in a home.

AIR: HOT AND COLD

Normal Body Temperature

The bodies of many animals can live through great changes in temperature. Toads, turtles, snakes, and their relatives grow warm or cold with their surroundings. The blood in their bodies can change many degrees without danger to their health. They are active when they are warm enough. They are quiet when they are cold, sleeping all winter in a *dormant condition*. There is no need for them to take exercise in cold weather. It does them no harm to have cold blood.

Even some mammals, such as bats, bears, and woodchucks, can live through wide changes of body temperatures. They, too, can be dormant during the winter season, although their blood does not become so cold as the blood of a reptile at such times.

The bodies of most mammals, however, are not able to change much. They cannot live if their blood becomes much cooler or much warmer than usual.

Such animals have natural ways of keeping well in summer. They may seek shady places and remain quiet during the middle of hot, sunny days. Some animals can sweat (perspire) if they exercise in the heat. The moisture on the surface of their bodies goes off in the air in the form of vapor; and this change helps keep them from becoming too warm.

We say that heat *evaporates* water (changes the liquid to a vapor or gas form). The heat that evaporates the drops of sweat on an animal's body is taken from the body itself. Thus the process of evaporation cools the body by removing some of its heat.

Some animals, like dogs, cannot sweat; but they can "cool off" by opening their mouths and letting moisture evaporate from their hot, wet tongues.

These mammals have natural ways, too, of keeping well in cold weather. Their fur coats are thicker in winter. They have snug shelters, too, in which to rest when they are not exercising.

A thermometer which is made to record the temperature of the human body usually has an arrow pointing to or near 98.6°. That is

Courtesy Taylor Instrument Companies

A thermometer to record body heat

called the *normal temperature*. People have blood about as warm as that when they are in good health. Indeed, in all your life, your blood will not become more than a few degrees cooler than normal. Neither will it be more than a few degrees warmer. You could not live through any such change as takes place in a woodchuck's blood.

Of course you do not need to worry about these matters. You do not need to keep a thermometer in your mouth to tell whether you are getting too hot or too cold. When you are in good health, the temperature of your body will take care of itself if you follow the three natural rules of the furry animals:

(1) Avoid too violent exercise in summer sunshine.

(2) Wear warmer outdoor body coverings in winter than in summer.

(3) Have a comfortable winter home shelter to enjoy when you are not exercising out of doors.

Hot Air for Winter

It is healthful to take walks or to play and run out of doors in cold weather. The low temperature of our surroundings does us no harm at such times if we are properly clothed. Our hearts beat faster when we exercise, the blood is pumped more rapidly through our bodies, and we may keep warm even in zero weather.

But when we have indoor clothing on and are quiet in the house, we are not comfortable unless the surrounding air is right. There should be some moisture in the air we breathe, though not too much. And, most important of all, there should be an ample supply of fresh air, with its life-giving oxygen.

Some people like to sit in rooms that feel too warm to others. In the United States most people prefer a house temperature of about 70°. Indeed, two degrees lower than this is marked "health temperature" on some thermometers. We should wear clothes that keep us warm enough, but not too warm, to be comfortable in rooms of about that temperature when we are sitting quietly indoors. It is a very good plan to have a thermometer in a living room so that the temperature may be known at any time. Then the air may be made warmer or cooler when it is not right.

How a Thermometer Works

A thermometer, as you know, is a device used to measure temperature. The word *thermometer* is made from two Greek words — *therme* meaning *heat* and *metron* meaning *measure*.

Galileo, an Italian scientist, is said to have invented the first thermometer about 1593; and thermometers of one kind or another have been used ever since that time. Galileo's thermometer, however, was not the kind you see in your home or at school. You doubtless look at a Fahrenheit thermometer to learn how warm a room is; so we shall tell you about that kind.

The common thermometer, for household use, is known as a *liquid thermometer*. An instrument of this sort consists of a slender glass tube with a bulb at the lower end. The bulb (and part of the tube) contains some liquid — usually mercury. As you have learned, nearly all substances expand, or become larger, when heated and contract, or become smaller, as they are cooled. Mercury is a liquid that expands and contracts nearly the same amount for each degree of change in temperature. That is one reason why it is a good liquid to use in thermometers. The mercury rises in the narrow tube when it is expanded by the heat, as that is the only empty space where it can go. It shrinks toward the bulb when it is contracted by the cold.

Gabriel Daniel Fahrenheit, a German scientist living in Holland, first used mercury in thermometers more than two hundred

years ago. He placed a tube with mercury in it in some salt and ice and a chemical containing ammonia. This mixture was so cold that the mercury shrank, or contracted, until it was very low. Fahrenheit called this lowest point to which the mercury sank in his tube *zero*. He marked this zero point on the glass tube 0°. Then he found the place to which the mercury rose in the tube at the heat of the human body. He marked this point also and finally called it 96°. (This point is marked at 98° on ordinary thermometers now in use. On a physician's thermometer it is of course marked near 98.6°.) Later the freezing point and the boiling point of water were found to be the most convenient fixed points. On the Fahrenheit scale these were found to be 32° and 212°. You may be interested to look at some thermometers to notice whether the zero point and freezing point, and some other points, are indicated by words at one side of the figures.   Is the temperature in the room in which you are sitting about 68° F. — the so-called health temperature?   (*F*. stands for *Fahrenheit*.)

Wood and Charcoal

Different kinds of fuel may be used to provide heat for houses.   Wood was the cheapest and most convenient fuel to use in most parts of our country before the forests were cut from so many places.   Some people still use wood to heat their homes.

Any kind of dry wood will burn, but each kind has a way of its own.   Some kinds blaze fiercely and burn quickly.   Others last a much longer time, giving out a more even heat all the while.

People speak of burning soft wood or hard wood.   White pine is an example of soft wood.   It has pitch (sap with resin) in it

and it burns rapidly with flaring flames.
Maple and beech are examples of hard wood.

Courtesy U. S. Forest Service

Wood that has been cut for fuel

A good chunk of one of these will keep burning all night if the drafts of the stove are cor-
rectly adjusted.

Charcoal, which is made from wood, is burned for fuel in some places. It is sold in small bags in Europe, where it is used more than it is in this country. In making quantities of charcoal, men pile wood in a special way, cover it with earth, and set it afire. The wood does not burn to ashes when it is covered in this way. The black charcoal that remains is collected and sold. Charcoal may also be prepared in an oven, called a *retort*. The wood does not burn in the retort; but it is heated in such a way that the moisture and gases are driven off and only the black *char*, or charred wood, is left.

You may wonder why we secure a better fuel by burning off part of the material that is in wood. Charcoal is nearly all *carbon*, which is the most important part of any fuel. While the gases driven from the wood in making charcoal have some value as fuel, it is the charcoal that is stronger as a heat producer.   Since there is no smoke, no soot, and very little flame in burning charcoal, this is a cleaner fuel than wood to burn. Not only can a hotter fire be made with it but it is also easier to control.

Coal and Coke

Coal is the chief fuel in parts of the world where it is found.   It is also carried to distant localities and used in many places where wood is no longer abundant.

This is a mineral fuel.   It is changed from ancient trees or other plants.   Ages ago, as you probably know, giant plants grew in swamps in certain places.   When they died, they fell into water and formed layers of decaying matter.   Gases were present in these substances under the water.   Some of these gases (*nitrogen, hydrogen, oxygen*) were given off in the process of decaying.   The chief substance that was left was a mass of carbon.

Such masses were pressed under heavy loads of sand and mud that settled on top of them. In places the earth's crust became heated by the pressure of shifting parts. This heat as well as the pressure helped change the masses of carbon.

Coal is not all exactly alike. It varies because of differences in the plant substance, the length of time it has been formed, and differences in pressure and heat.

Anthracite, or hard coal, is the pleasantest kind to use. It burns longer and with a steadier heat, and it gives off very little smoke. It is broken at the mines into rather small pieces that are very black and shiny.

Bituminous coal, or soft coal, burns with a bright yellow flame. It is the cheapest and commonest coal to be used as fuel in most places. It comes from the mine in big lumps. If it is not properly handled, it may give off a great deal of dirty black smoke as it burns. Such smoke pours out of chim-

Courtesy U. S. Bureau of Mines

Drilling hard coal in a mine

neys in dark clouds. Its sticky soot settles on city buildings, making them untidy. It injures growing trees and other plants by stopping up the pores in their leaves, through which they need to get air. It makes the air unpleasant for people to breathe and often unhealthful. Some cities are finding ways to lessen what is called the smoke nuisance. They are using better methods in burning soft coal.

Lignite, or brown coal, is even softer than bituminous coal. It has not undergone so complete a change as the two other kinds of coal mentioned. It is possible to find woody parts in it. It is more like peat than are other kinds of coal.

Peat is not really coal. It might be called "half-made coal." It is found as decayed masses of plant matter in many swampy places. It is used as a common fuel in Ireland and some other countries.

Coke is made from coal as charcoal is made from wood. Coal, usually soft coal, is heated in great retorts. The hot coal gives off gas, and the coke remains. Just as charcoal is a better fuel for many purposes than is wood, from which it is made, so is coke superior to coal in some ways. Coke can be used in some industries where coal could not be used because the gases and smoke given off from the burning coal would interfere seriously with what is being made. Where great heat is needed, as in the manufacture of iron and steel, coke is better than coal for fuel.

Gas

The gas which is given off by the heated coal in the retorts is cleaned and burned for the heat it gives. If it is not "rich" enough to give good heat, oil may be added to make it better.

Natural gas may also be used for fuel.

This is not obtained from coal heated in retorts. It is taken from gas wells, and it is also derived from petroleum wells.

Oil

Petroleum, like natural gas, is taken from wells. It has various other names — some of them are mineral oil, earth oil, and crude oil. It is purified and separated into different oils, one of which is gasoline and another kerosene. People use petroleum in many ways. It serves as a fuel oil in some buildings.

Electricity

You have read, in another chapter, that a current of electricity will heat any wire along which it is difficult for it to pass. Such heat is used in an electric iron or a toaster that is provided with a coil of wire through which the current must go. This coil becomes bright red and hot. Electricity is used in this same way, though on a larger scale, to heat rooms or even entire houses.

Courtesy U. S. Bureau of Mines

Oklahoma oil wells, with storage tanks and derricks

Stoves and Furnaces

In the early days of our country people did not have stoves. They burned wood in *fireplaces*. A fireplace is a bricked space leading into a chimney. It is pleasant to watch the flames flickering in a fireplace and to see the glowing coals. Such an indoor fire may be enjoyed like an outdoor bonfire.

Many people still have fireplaces so that they may watch the open fire. The air in the room, however, cannot be heated evenly in this way. The room is too hot in front of the fire and too cool in other places. So, in cold climates, stoves or furnaces are needed for heating purposes.

About two hundred years ago a man named Benjamin Franklin invented a stove. It had an open front and a frame that was shaped much like a fireplace. At the back there was a pipe that led into a chimney. A chunk of burning wood lasts longer in a

Franklin stove than it does in a fireplace, and the air in the room is heated more evenly.

Benjamin Franklin was a generous man. He permitted anyone to make this kind of

Courtesy Wood and Bishop, Bangor, Maine

A Franklin stove with the doors open

stove who wished to do so. It was possible to make Franklin stoves at low cost. Even poor people could buy them.

Since Franklin's time, many sorts of stoves and furnaces have been invented to give more and better-controlled heat. Most of them are made of iron with rather large doors that can be opened when it is necessary to put in fuel or remove ashes.

These stoves and furnaces also have small doors, or metal plates, that move by sliding or swinging when they are opened to let air into the stove or furnace. How fast a fire burns depends in part on the amount of air admitted to it. This is controlled by these small doors, or metal plates. When the door (marked A in the drawing on page 200) beneath the fire is open and the one (marked B) in the smoke pipe, or flue, is closed, the rush of hot gas and smoke from the fire goes up the chimney and reduces the air pressure* in the chimney. Cold air entering the fire box from the outside is pushed through the fuel in the fire box. The fuel then burns vigorously,

---

* Weight of air and air pressure are mentioned in another chapter in this book.

and the fire becomes very hot.　If we close the door (*A*) that admits air in front of the fuel and open the one (*B*) that is in the flue, or smoke pipe, then the air takes a short cut

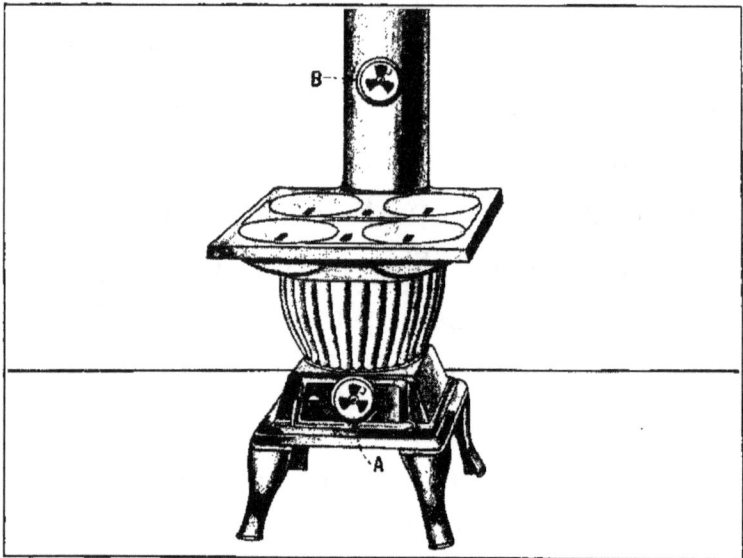

Notice the metal plates (*A* and *B*) that may be opened or closed.

into the smoke pipe and up the hot chimney without ever going through the fire.　The cheated and hungry fire now burns more and more slowly and may even go out.

For many years wood stoves and coal stoves were placed directly in the rooms to be heated. But they take considerable space and are often troublesome in other ways. So in most city and in many country homes one furnace is now used instead of two or more smaller stoves. A furnace for heating by hot air is really a big stove surrounded by a jacket, or casing. It is usually placed in the basement or cellar.

Pipes lead from the hood of the furnace to the different rooms that are to be heated. Furnace pipes may be large. One or more return air pipes may carry cold air down to the bottom of the furnace casing, and others carry currents of hot air from the hood to the different rooms. The hot-air pipes lead into *registers*, or openings, in the floors or the walls of the rooms.

Or the pipes may be small pipes to carry currents of water instead of air. The heated water rises in the pipes to the floors above, where it passes through *radiators*, or sets of pipes. The cold water in the system is

A furnace for heating by hot air

forced to return to the furnace to be reheated.

In many buildings the water in the furnace pipes is heated until it becomes steam. Then the steam goes up through the pipes to radiators in the rooms. In time as the steam in the radiators cools, it changes from vapor of course to liquid water. This water is carried through pipes back to the furnace to be reheated in the boiler there.

Ventilation

Any fire, to burn, must have air (with its oxygen), just as we must have air to breathe. If we close a room tightly and have a fire, the oxygen in the air will soon be used. A person in such a room may become drowsy or ill.

In order to prevent trouble of this sort, a house should have proper ventilation. By ventilation, you may remember, we mean ways of letting fresh air into the house so that there may be enough for the people and for the fires. Many large buildings have special air shafts for this purpose. A simple way to ventilate a "stuffy" room in cold weather is to lower a window a little from the top.

Comfortable Air in Houses

It is possible to provide houses with the kind of indoor weather that is most comfortable at all times of the year — cool air in summer as well as warm air in winter. It is especially important that the moisture in the air should be right. Some buildings are equipped with machines that keep the air in a good condition — cool enough and with the proper amount of moisture. Preparing this kind of air is called *air conditioning.*

Did you ever go into a theater or restaurant or other public building in hot summer weather and find the air cooler and pleasanter than it was on the street?

Cooling by Evaporation

On page 179 you read that the evaporation of moisture from the surface of an animal's body helps cool the body. There are pores in the skin through which this moisture can pass.

People have learned how to keep water cool without ice, by allowing some of it to evaporate from the surface of the container. In India it is a common practice to use porous water jars through the sides of which droplets of the water may seep. This seeping water evaporates from the surface of the jar. Heat used in this evaporation is taken from the jar, and thus the jar is cooled and so is the water it contains. Army water coolers that work in this manner are made for the use of soldiers.

Questions and Activities

In this chapter you have read about several different kinds of fuels. Name one fuel that is a gas, one that is a liquid, and two fuels that are solids.

What is meant by a *dormant* plant or animal?
If you do not know, read Chapters Ten and Eleven
in the book *Through Four Seasons.**

Did a nurse or physician ever put a thermometer
into your mouth? What reasons may there be for
doing so? A thermometer made for this purpose
is not quite like one made to record the temperature
in a room or out of doors. Explain what some of
the differences are. (Look at real thermometers
or at the pictures on pages 180 and 185.)

Choose one of the following kinds of stoves or
furnaces to study: one that burns coal or gas or
oil or wood for fuel. Choose a kind that it is
possible for you to visit in some home or other
building.

Each kind of flame uses oxygen and gives off
gases that are unpleasant (sometimes dangerous)
to breathe. Find out how the fresh air reaches
the fuel in the kind of stove you choose to study.
Find out what becomes of the bad air from the
flame.

Does hot air rise or fall? Could a house be
heated by a furnace in a room on the top floor
and a chimney leading down to the ground?

*See page 424 for book list.

Explain how your schoolroom is ventilated.

An ordinary unglazed flowerpot is porous enough to let moisture seep through it. Take such a pot and close the hole in the bottom with a cork. Take also a glass jar that will hold about the same amount of water as the flowerpot. The glass jar is not porous. Fill both containers with water of the same temperature and put covers over their tops. Keep them for a few hours in a warm room or out of doors in the shade on a warm, sunny day. In which container does the water remain cooler? Why?

You may find drops of moisture on the outside of the glass jar. If so, these do not come through the sides of the jar. How do they get there? If you have forgotten, read pages 180 and 181 of *Surprises*.

On pages 92 and 93 you read about building boards, made of different materials called nonconductors, which are used in the walls of some houses as heat insulators. It is appropriate to mention again the subject of heat insulation in connection with this chapter. The space between the studs, or light timbers, in the walls of a house may be filled with such materials as sawdust, ground cork, and rock wool.

Rock wool looks much like wool that is shorn from sheep. It is made by melting rocks of certain kinds and by blowing the hot material out into fine fibers.

Imagine two wooden houses in one of our northern states — one with no insulating material between the studs of the outer walls and one with rock wool (or other insulating material) between the studs. Which house would require more fuel to keep the temperature of the rooms at about 68° F. during the winter? Explain why.

HOME AS A WORKSHOP

PART FOUR

CARPENTER TOOLS

You are quite likely to find carpenter tools on a shelf or in a chest somewhere about the house even if there is no carpenter in your family. Perhaps you often use some of the common tools yourself. If so, you have doubtless admired the ease with which you can do a bit of work that would be difficult, if not impossible, for you to do without the help of the proper tool.

There are very good reasons for the choice of the materials used in tools and good reasons, too, for the shapes of these different implements. We think it may add to your interest in tools and to your pleasure in using them to know why and how some of them work as they do.

Claw Hammer for Pounding and Pulling

No one would be likely to make claw-hammer heads of aluminum. This material would not be heavy enough or hard enough or strong enough for the purpose. Very cheap hammers have been made with heads of cast iron. Now cast iron does not lack in weight or in hardness, but it does lack in strength. Any claws of an aluminum hammer head would certainly become bent during ordinary use; but the claws of a cast-iron head are about equally useless, for they are too brittle for the severe strain of prying. If you look through a catalog in which proper tools are listed, you will find the statement that the hammers are made with steel heads. Steel is, indeed, hard and strong and heavy enough for all the work that a hammer is usually expected to do.

When you swing a hammer to drive a nail, the effect of the blow depends not only on how big and heavy is the head of the hammer, but also on how hard you strike with it. We may call the quantity of matter in the hammer head its *mass*. How hard you strike depends upon how fast the hammer is swung through the air. We may call that speed *velocity*. When something having mass, like the hammer head, is put into motion to give velocity, it is said to acquire *momentum*. We can measure this momentum and have some idea about how hard the hammer blow was struck. Momentum equals mass multiplied by velocity.

Suppose that your hammer head has a mass of two pounds and suppose that you swing it with a speed, or velocity, of ten feet a second. Then you would hit the nail with a momentum of twenty, because 2 (mass) × 10 (velocity) = 20 (momentum).

If you took another hammer having a head with a mass of one pound, you would need to move it twice as fast, or at the rate of twenty feet a second, to bang the nail with the same momentum, because 1 (mass) × 20 (velocity) = 20 (momentum).

But, if you had a hammer head as heavy as that of a four-pound ax head, then you would need to swing your hammer only half as fast as you did your two-pound hammer head in order to hit the nail with the same momentum, since 4 (mass) × 5 (velocity) = 20 (momentum).

The claws of the hammer are of no help to you in striking a blow. But of course, if a nail bends when you hit it, you are glad to have at hand some means of pulling the bent nail. You do not need to lay your hammer aside and seek another tool, for your hammer itself is really two tools — a pounding tool and a pulling tool.

When you pull a nail with the claws of a hammer, you are using one kind of lever.

On page 140 of this book you read that a well sweep is a lever. Levers are of different kinds. They are used in the simplest as well as in the most complicated machines. The law by

A seesaw is a lever.

which they all work, however, is illustrated by a seesaw.

On a seesaw, as you know, any small boy, by sitting farther from the turning point, or *fulcrum*, can counterbalance a larger boy.

Thus a boy weighing fifty pounds, by sitting eight feet from the pivot, may hold his own with a hundred-pound boy who sits four feet from it. You may put the lever law in these words: "*Working force* times its distance from fulcrum equals *resisting force* times its distance from fulcrum." This law may sound a bit hard to you at first; but, if you will look carefully at the drawing on page 215, you will see that it is really rather easy to understand. If you call the smaller boy Working Force and the larger boy Resisting Force, you may say that 50 (Working Force) × 8 (Working Force's distance from fulcrum), which is 400 = 100 (Resisting Force) × 4 (Resisting Force's distance from fulcrum), which is 400.

How does a claw hammer resemble a see-saw? Hold a claw hammer on a board in position to pull a nail that has been pounded for part of its length into the board. Call the pull you give with your hand "working force." Call the force it will take to pull the nail free from the board "resisting force." The hammer of course is the lever. It is a

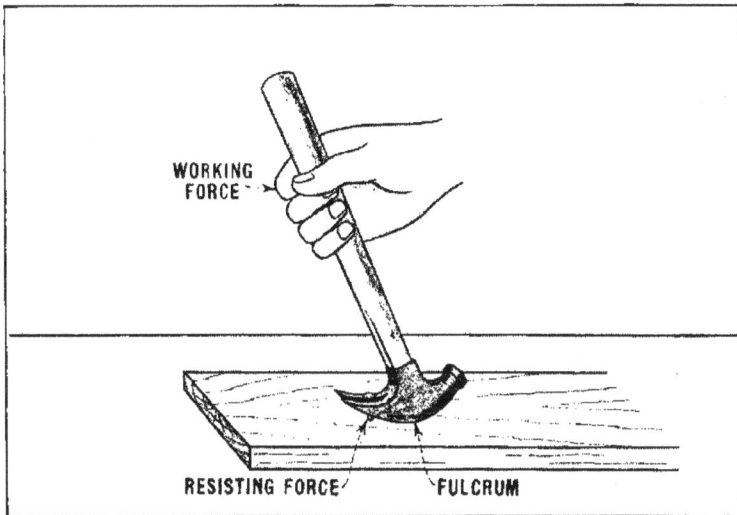

WORKING FORCE

RESISTING FORCE   FULCRUM

A hammer used as a lever

rather crooked lever and not a straight one like a seesaw. The turning point, or fulcrum, is the place where the hammer touches the board.

Suppose that the head of the nail is two inches from the fulcrum, and that you grasp the handle of the hammer at a place eight inches from the fulcrum. Then, if you give a twenty-five-pound pull with your hand, you can exert a 100-pound pull upon the nailhead. This must be so because, according to the lever law, working force times its distance equals resisting force times its distance. In other words, 25 (working force or the pull of your hand) × 8 (working force's distance), which is 200 = 100 (resisting force or hold of the nail on the board) × 2 (resisting force's distance), which is 200.

Snips or Shears

Perhaps some day you may wish to cut a piece from a sheet of tin. Then you will be glad to find a pair of tinners' snips on the tool shelf. Tinners' snips, as you might guess from the name, are shears for cutting metal. To be sure, you could cut paper and cloth with them; but, as they are awkward to handle and heavy, you would not care to use them for that purpose. This tool, too, works according to the lever law.

Thus, if you grasp the handles at a place eight inches from the turning point, or fulcrum, with a five-pound force, you exert

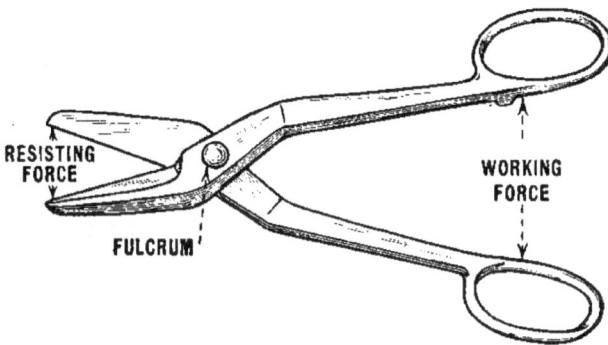

Tinners' snips work by the lever law.

twenty pounds' cutting pressure on a piece of tin inserted between the blades at two inches from the fulcrum. You can tell that this is so because 5 (working force) × 8 (working force's distance), which is 40 = 20 (resisting force) × 2 (resisting force's distance), which is 40. But, if the piece of tin is placed four inches from the fulcrum, then your pressure on the tin would need to be only ten pounds. You may have noticed that it is easier to cut a piece of thick pasteboard with an ordinary pair of shears when the pasteboard is near the fulcrum than it is to cut the pasteboard with the tips of the blades. Now that you know what the lever law is, you can see why this is so.

The well sweep, the seesaw, the claw hammer, the tinners' snips (and common shears and scissors) are all alike in one respect. In each of these, the fulcrum is between the working force and the resisting force. Levers of this sort are known as levers of the *first class*. There are two other classes of levers, of which we shall speak in the next chapter.

Wood-cutting Tools and Chips

There are various tools for cutting wood, in one way or another, that remove parts of the wood worked upon in the form of chips.

If you cut a stick in two with a jackknife, you are likely to slant your strokes first in one direction and then in the other so that you make a V-notch in the stick.

A man chopping through a log swings his ax first from the right and then from the left so that he cuts out wedge-shaped chips and makes a V-notch in the log. With his alternate strokes he forms an opening wider at the top and can conveniently chop deeper and deeper into the log.

Such a boring tool as a gimlet or a bit, with a screw-shaped cutting blade, is smallest at the point which first enters the wood. The spiral blade of either one of these tools is so shaped as to leave open winding ways from the bottom to the top of the hole it bores. Through these open winding paths the chips can find an upward outlet and so pass out of the way of the tool.

Instead of having one large blade, like an ax, a saw has a series of small blades (called teeth) arranged in a row — or rather in two rows. No doubt you have noticed that the teeth of a saw, unlike those of a toothed bread knife, are bent alternately in opposite directions. The process of so bending, or inclining, the teeth is called *setting* the saw.

Because of this setting, or alternate bending, of the teeth, the toothed edge of the saw sticks out beyond the rest of the blade on each side. This enables the tool to cut a *kerf* (channel) that is wider than the thickness of the saw blade. Thus, if the teeth are properly set, the kerf is wide enough so that the saw blade can be moved back and forth without too much friction (rubbing). More friction of course would mean more work in moving the saw.

Since the two rows of bent teeth, taken together, are wider than the rest of the blade, we might think that the saw thrusts its thickest part into the wood first. However, we must not forget that each tooth is a cutting blade and that each pointed tooth is so sharpened that its thinnest edge enters the wood first. In this respect, then, each tooth of the saw is like a knife or an ax. You can see, too, that the oppositely inclined teeth of the saw work in the wood in somewhat the same way as the alternately inclined strokes of an ax or hatchet in chopping a notch — hitting first from one side and then from the other.

All sorts of wood-cutting tools must of course provide an outlet for their chips. The wide V-opening cut by the ax gives a free escape for the large wedge-shaped chips. As we have remarked before, the curving open channels of the gimlet and bit are upward paths for the chips of such tools. How does a saw get rid of its chips? Watch one, as it moves

back and forth through a piece of wood, and
notice that the very small chips cut by this
tool are dragged out of the kerf between the
teeth of the saw in the form of sawdust.

Some Screws

An instrument with a screw motion is often
used when great force is needed. We have
already mentioned two boring tools, the gimlet and bit, which cut with a screw motion.
In the next chapter we shall speak of certain
screw devices that are used in kitchens. Just
now we may turn our attention to some of the
twisting tools a carpenter finds useful.

The metal *wood screws*, for instance, with
which a hinge is fastened to a cupboard door,
will hold the hinge in place much more firmly
(with much more force) than would nails of
the same length. The weight and motion
of the door, as it swings to open and close,
would tend to pull the nails and loosen the

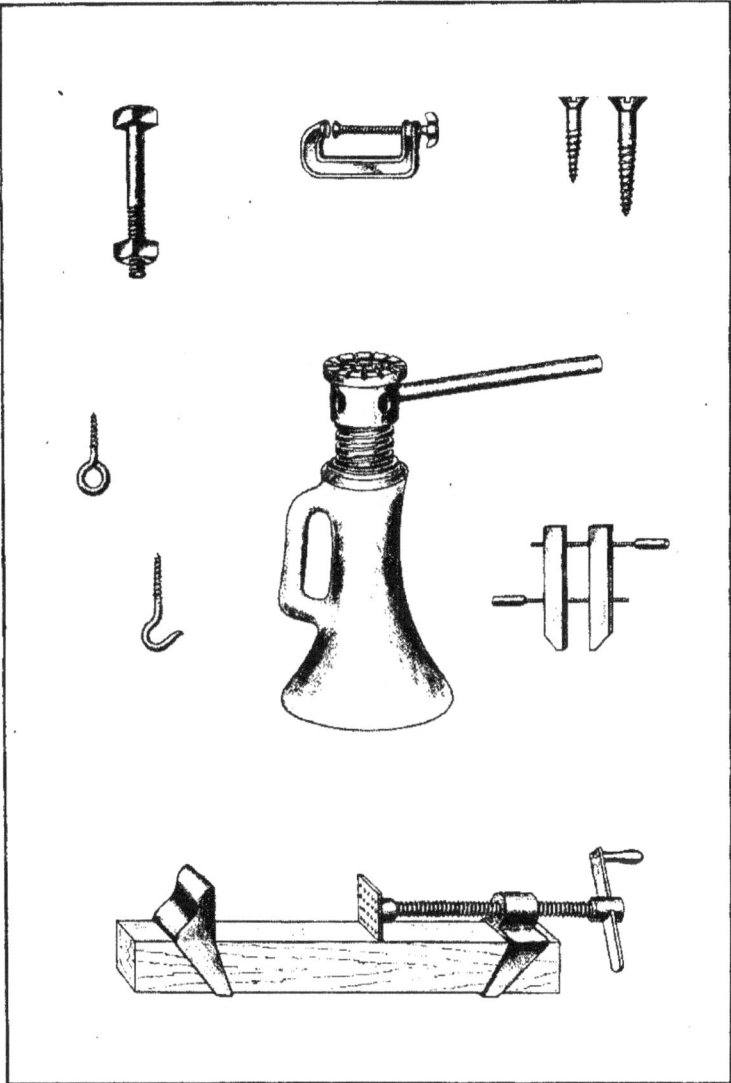

Some different forms of screws and screw tools

hinge. A nail, having a smooth surface, is somewhat easily pulled. The spiral part of the screw, however, catches hold of the wood into which it is turned and can, with greater force, keep the hinge in place. This spiral part of a screw is called its *thread*. (The word *thread* means *that which is twisted*.)

Metal rods called *bolts* are used to hold objects together. The screw thread at one end of a bolt fits into a *nut* — a metal block having a round central hole large enough to receive the bolt. There is a screw thread on the inside of the nut. When the nut is twisted around the end of the bolt, the two sets of screw threads fit in such a way as to keep the nut from being pulled off. The objects held between the nut and the head of the bolt are thus pressed together and kept in place.

Clamps and *vises* are screw-motioned tools used to hold pieces of wood or other objects steady while they are being worked upon.

The *jackscrew* is a lifting screw. With this tool it is possible to raise a heavy weight by using a small force. You may think, then, that the work of a jackscrew is similar to that of a lever. So it is. It works, indeed, in accordance with the same "law of machines" as does a lever. However, the jackscrew is more complicated than a simple lever and you will doubtless prefer not to read about jackscrew mathematics in this book.

QUESTIONS AND ACTIVITIES

Look at the picture of a well sweep on page 140 of this book or in some other book. Then draw a picture of a well sweep on a piece of paper. On your picture label the fulcrum, the place where the working force operates, and the place where the resisting force, or weight, is.

Use a twelve-inch ruler for a toy seesaw. Rest it on the edge of a box or something else you can use for a fulcrum. Have one half of the ruler on each side of the fulcrum. Take two objects of exactly the same weight. Call one object "working force" and the other "resisting force." Fasten "resisting force" to the ruler five inches from the fulcrum. You may do this by tying "resisting force" so that it hangs from the under side. How many inches from the fulcrum will "working force" need to be in order to balance "resisting force"?

When you see tools in your own home or in the home of some friend or in a hardware store, try to see how many of them work as levers and how many have screw parts.

In this chapter we have tried to make clear the fact that a lever is a means of multiplying force. If you know anyone who is willing to lend you a jack such as is often used to raise an automobile for the purpose of changing a tire, you may like to make a test for yourself to see that you can raise with the help of the jack some object that is altogether too heavy for you to lift without some such help. You can doubtless think of tricks you can do with a crowbar or some other implement used as a lever.

A COOK'S TOOLS

The Name *Kitchen*

A kitchen is a *cooking room*. The English word *kitchen* was changed from a Latin word. This Latin word is *coquina*. The people in many countries have a word meaning "cooking room" that was changed from this same Latin word. If you learn to speak Danish or French or German or Italian or certain other languages, you will learn several ways of changing the Latin word *coquina*. There is no room in the home that is more important than the one where cooking is done. So of course people in different countries need names for it.

You know about some of the tools that masons, carpenters, plumbers, and certain other craftsmen use. The tools, or devices, found in a kitchen are quite as important to a cook as a hammer and a saw are to a carpenter. One good way to become acquainted

Photo by J. C. Allen, from R. I. Nesmith and Associates

Busy with kitchen tools

with such tools is to watch a cook at work. An even better way is to use them yourself while you learn to prepare some kinds of food.

Measuring Devices

Suppose that a cook were making four fruit cakes and wished to mix them at the same time. She might turn to a recipe in her cookbook that stated the various materials needed. Some of them might be the following: 1 pound sifted flour, 1 teaspoon baking powder, $\frac{1}{2}$ teaspoon cloves, $\frac{1}{2}$ teaspoon cinnamon, 1 pound butter, 1 pound sugar, 10 beaten eggs, $\frac{1}{2}$ pound candied pineapple, 1 pound seeded dates, 1 pound seedless raisins, $\frac{1}{2}$ pound chopped nut meats, 1 cup honey, 1 cup molasses, $\frac{1}{2}$ cup sweet cider. The cookbook might include these instructions: "Bake in four pans ($7 \times 9 \times 2$ inches) in a slow oven ($250°$ F.) for $3\frac{1}{4}$ hours." What measuring devices would the cook need in order to make her fruit cakes?

She would need a device for measuring *weight*. On a household balance, or scale, she could measure her pound of sifted flour

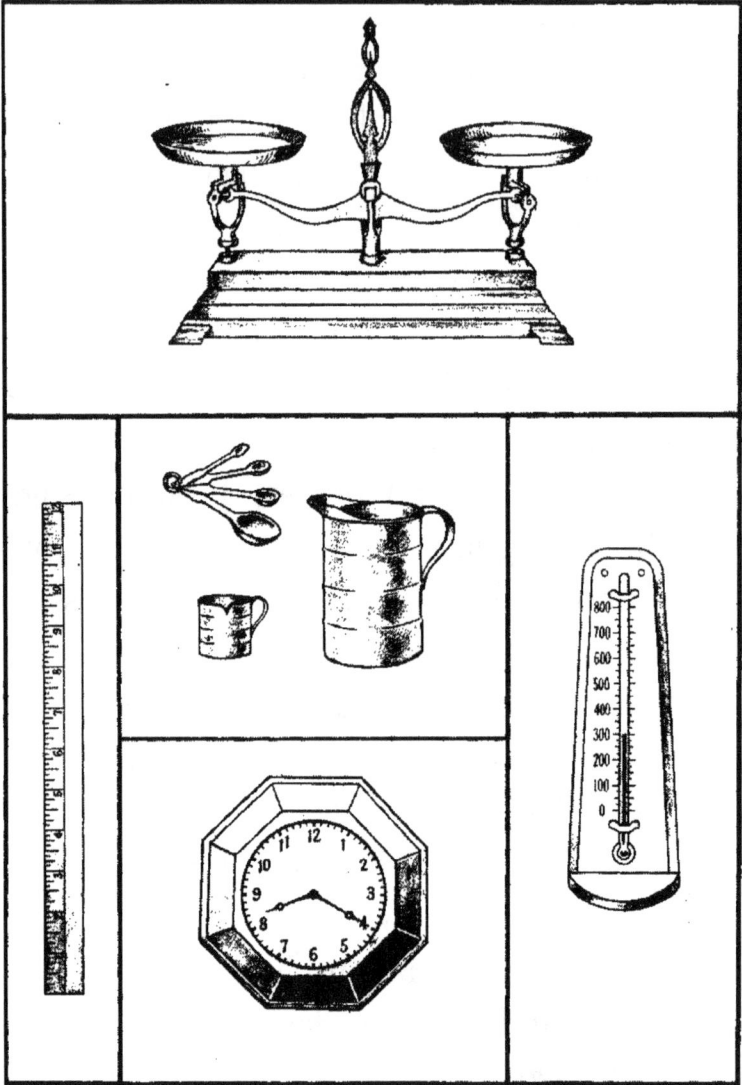

Measuring devices that are useful in a kitchen

and the required amounts of other materials indicated by weight.

Devices for measuring *volume* (bulk) would be needed to make sure that she used the right amounts of baking powder, honey, and other materials indicated by spoons and cups.

In order to be certain that her pans were large enough and not too large, the cook would need a device for measuring *length* — such as a tape or a rule (ruler).

In order to measure the *temperature* of the oven, the cook would need to look at a thermometer. There are oven thermometers that indicate the heat from 200° to 600° F. Some are made to be placed inside the oven, and some are permanently set in a little frame in the oven door.

Of course a clock or a watch would be needed to measure the *time* allowed for baking the cakes.

Three Classes of Levers

On page 220 you read that levers of the first class have the fulcrum between the working force and the resisting force. You will see, by looking at the picture, that the balance shown on page 232 is a lever of this class.

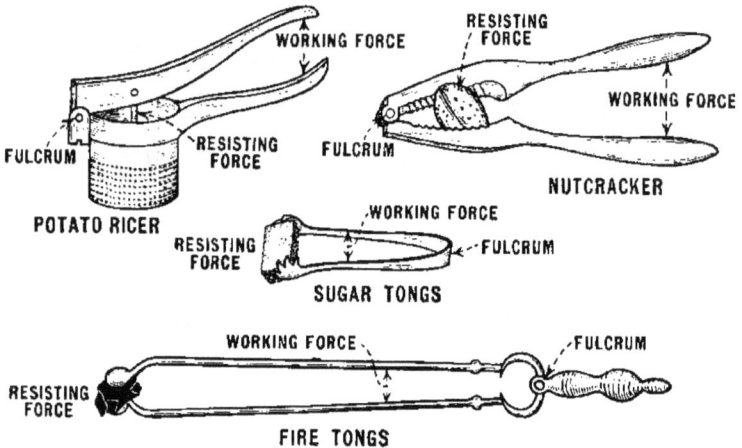

WORKING FORCE

RESISTING FORCE

FULCRUM

RESISTING FORCE

POTATO RICER

RESISTING FORCE

FULCRUM

WORKING FORCE

NUTCRACKER

WORKING FORCE

RESISTING FORCE

FULCRUM

SUGAR TONGS

WORKING FORCE

FULCRUM

RESISTING FORCE

FIRE TONGS

Levers of the second and third classes

When you use a nutcracker or a potato ricer, you have the help of a lever of the *second class.* In these tools the resisting force is between the fulcrum and the working force.

The working force is applied between the fulcrum and the resisting force in levers of the *third class*. You may notice that you are using a tool of this sort the next time you move a lump of sugar with sugar tongs or lift a bit of burning coal or other object with fire tongs.

Screws Again

In another chapter you read about screws and the force that can be exerted by their motion. You cannot spend much time in most kitchens without meeting a screw that is useful in one way or another. Perhaps the first one you may notice will be a corkscrew. Every time you twist the handle of a water faucet you turn a screw. Some fruit jars and certain other bottles and cans have screw covers. The rims of such containers have screw threads and the covers fit over these as a nut fits over a bolt. If you fasten a

meat chopper to the edge of a table, you do
so with a screw clamp. Then, when you turn
the crank of this same tool, you move a

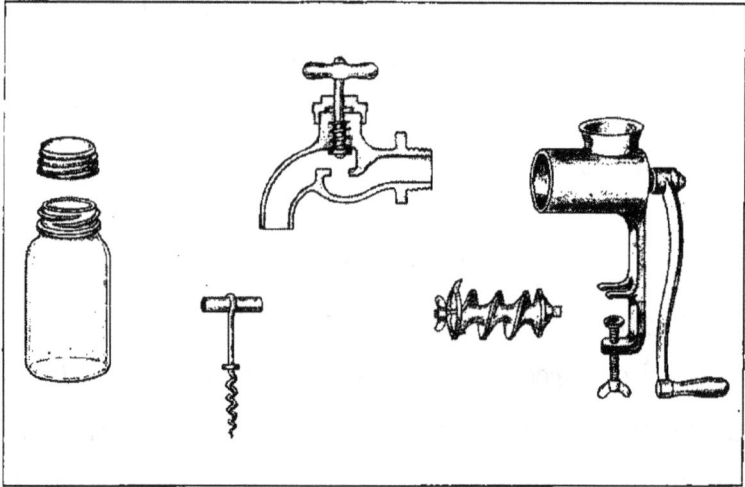

Some kitchen screws

tapered cutting screw that minces the meat
and presses it out through the opening.

Cranks

We have just mentioned the *crank* of a
meat chopper. This is the *arm* attached to
the end of the cutting *shaft*, or bar, that

turns around in the machine. There are various sorts of cranks used in more complicated machinery, but most of those found in a kitchen work in much the same manner as that of the meat chopper. Thus a bread mixer or an ice-cream freezer is worked with a crank. So is a grate shaker in a coal range, and a hand-turned clothes wringer.

You may think that the cranks of these devices resemble the lever arm of the jack-screw of which you read in another chapter. Of course they do. In fact, you may call any one of these cranks the working-force arm of a lever if you like. Perhaps, too, you may be interested to see for yourself what part of a meat chopper or a bread mixer you might call the fulcrum and where the resisting force is located. These cranks, as you will learn if you turn them, give us added reasons to be thankful for levers, by means of which heavy work can be done with little effort.

The word *crank* means *bend, twist, wind, turn*. You can readily see why this word is a good one to give to a mechanical arm that twists or turns a shaft. You may also be interested to think why the same word may be applied to a person who has whims — peculiar twists or turns in his mind.

Dishes to Stand Heat and Cold

Since some food is frozen and some is cooked, the dishes that contain it must be of materials that will not be harmed by being made very hot or very cold.

You may have noticed that some materials cool much more slowly than others. We say that they hold the heat. Crockery made of selected clay is often used when food is being baked slowly for a number of hours. A clay bean pot is such a dish, and you may have noticed that food cooked in a bean pot remains hot for a rather long time after it is removed from the oven.

Common sorts of glass and china are likely to crack when heated in an oven. They may actually break or may become full of tiny cracks. Such cracking is called *crazing*.

Chemists have preferred to use thin glass *test tubes* in which to boil liquids. Such thin wares would not last long in a kitchen; but thick dishes, if made from certain kinds of glass, may be used for baking without danger of cracking.

Silica ware, made by melting pure sand, is also used by chemists. This is very tough and stands being taken from a hot oven and plunged directly into water. It is too costly for household use.

For some sorts of cooking, a dish is needed that may be heated very rapidly without harming it. Pans and skillets of cast iron, steel, or *aluminum* are often chosen for such service.

Since chemists have learned how to make aluminum at a moderate cost, this material is used for kitchen and many other purposes. Aluminum does not crack, as some other substances do, and heat passes rapidly through it to the food it contains.

Discolored Metals

Iron, as you know, rusts quickly if moist. In fact, cooked food should not be left standing long in any metal dish, as both the food and the dish may become discolored even if not otherwise injured. Mixtures of certain metals with iron, however, are not easily rusted and are not readily attacked by fruit juices or other vegetable acids. Mixtures of metals, you remember, are known as alloys.

Stainless steel is a name of some kinds of alloys that are easy to keep clean. Many knives are now made of stainless steel. It is not very remarkable that these alloys are lasting, since they contain such metals as nickel and chromium, which do not rust in the air and are not affected by mild acids. They may remain too costly for ordinary kitchen utensils, however. Their price is high mainly because the ores are hard to find and because it is expensive to get metals from them.

But some alloys of iron, such as ordinary steel, do rust if exposed to moisture. You may have seen garden spades and rakes covered with a bright brown powdery crust if they were left about the yard in damp weather. Some kitchen knives and forks become similarly discolored if they are allowed to remain wet. As you learned on page 102, rust, or iron oxide, is the substance that is formed when oxygen combines with iron.

You may often hear some one remark, "This silverware should be polished before it is used again." And sure enough, if you look at a spoon or another piece of silver, you may notice that sometimes it is not bright and pretty but that its surface is dull and dark colored. The tarnish on the silverware is a substance that is part silver and part sulfur. There are very small quantities of sulfur in many of our foods, and the yolk of an egg has considerable sulfur in it. You can doubtless guess why you are most likely to see a coat of such tarnish on silverware with which cooked egg has been eaten — especially if it has not been washed promptly after the meal. Rubbing with fine powder or paste made from fine powder such as chalk dust (whiting) removes the tarnish but will not scratch the silver under the tarnish.

If a copper dish is used for cooking or serving food, it should not be set aside with any of the food in it. Certain acids may combine with the copper and form a poison. The acid that is in vinegar will do this. This acid combines with copper and forms green crystals of a substance called *verdigris*. Verdigris is only one of the poisons that may be formed by copper and acids. You can guess what might happen if pickles or apple sauce were left for a few hours or days in a copper dish and then eaten.

It is of course important that copperware should be washed and dried soon after it is used. It is also well to polish it with some cleansing powder to keep it smooth and bright. Some people even think it is a good plan not to use dishes of copper for food at all — but that they may best be kept with antiques, merely for their beauty.

QUESTIONS AND ACTIVITIES

Prepare an exhibit with pictures of dishes and other utensils you would like to have in the kitchen if you were a cook. You may find such pictures in catalogs and on the advertising pages of papers and magazines.

Write a short essay about your exhibit. State a use to which each article may be put. If it works as a lever does or with a screw motion, say so.

Place a copper penny or a bit of copper wire in a saucer with a little vinegar. How soon do you find crystals of verdigris? Remember that this is a poison. Put it where it can do no harm.

Look for the word *verdigris* in an encyclopedia to find an important use for this substance.

There is an old French word *verd* meaning *green*. Do you see why a word with this meaning should be used in a name for the substance formed by copper and vinegar? Look in a dictionary for the words *verdant* and *verdure*. What do these words mean?

The balance shown on page 232 is a kind that is called an *equal-arm beam balance*. This is a reliable sort of balance. You will find beam balances in the laboratories of scientists who need to be exact in the amounts of materials they weigh for their experiments or other work.

If you look at common household instruments for weighing, however, you are likely to notice some that are *spring scales* instead of beam bal-

ances. As the name indicates, such a scale is
equipped with a spring that is coiled. Any weight
placed on the platform of the scales stretches this
spring. If 2 pounds stretch it 1 inch, then 4
pounds will stretch it 2 inches and 1 pound will
stretch it $\frac{1}{2}$ inch. A spring scale is fitted with a
pointer that moves over a scale face with figures to
show about how heavy an object is being weighed.
Spring scales are convenient to use, and they indicate weight closely enough for most household
purposes. They are not, however, as exact as
good beam balances.

You may be interested to look at balances and
scales you find in use in different places and to
notice whether they are beam balances or spring
scales.

Name two acids that are mentioned in the
chapter you have just read. (Look on pages 240
and 242 if you do not remember.) Another kind
of acid is mentioned on pages 32–33 of *Through
Four Seasons.** Read what is said about it.

* See page 424 for book list.

Using a vacuum cleaner -- a modern device

MATERIALS AND DEVICES FOR CLEANING

You do not need to be told that the rooms in a house would be unattractive places if they were not kept clean. You do not care to look at dust and rubbish. You do not like the odor of dirty objects. Besides being unpleasant to see and smell, such surroundings would become unsanitary.

Too much dust in the air you breathe may easily irritate the membrane (called *mucous membrane*) that lines your nose and throat. Then, too, as you may have learned from the book *Through Four Seasons*, much of the dust that floats in the air and settles on objects indoors and outdoors is live dust — such as bacteria and spores of molds and other fungi.

If bits of food are left on dishes, they may become sour or decayed. It is as important to keep dishes free from such substances as it is to keep our teeth clean. Molds and bacteria grow in spoiled food. Some of these are harmless, but others cause sickness if through carelessness they are allowed to come in contact with food that is later to be eaten.

Since it is important to have our daily surroundings as sanitary as possible, much attention has been given to the materials and devices that are useful for cleaning purposes in the kitchen and other rooms of the house.

Soap

Soap and water are the cleaning materials that are used in greatest amounts in the kitchen. A greasy plate washed in hot soapsuds and rinsed in boiling water may be allowed to drain until it is dry or it may be wiped with a clean cloth. Then it is ready for the shelf or to be used again.

In the days of your great-great-grand-mothers nearly all women made their own soap. They saved the grease from fat meat which they cooked. They saved, too, the ashes from the fireplace, where wood was burned.

There is *potassium* in wood ashes. The women put the ashes in water and dissolved the potassium compound. Then they boiled the potassium solution and grease together to make *soft soap*. This soap was too thin to be made into stiff cakes. It had to be dipped with spoons or cups. Perhaps you have never seen soft soap, but it was once the most common sort to be used for washing dishes and clothes.

When the women of those days wished to have *hard soap*, they stirred salt into the hot soft soap. There is *sodium* in salt. This sodium caused a change to take place in the soft mixture so that it became stiff and firm enough to be cut into cakes when it was cold and taken out of the kettle.

Today some soap is still made in homes, but most soap is prepared in factories under the direction of chemists. Enormous kettles, holding thousands of pounds, take the place of the black kettles women used to place over open fires in the farmyard while they made their soap.

But the learned chemists of today use the same three necessary substances that your great-great-grandmothers used years ago — oil (or grease), potassium compounds, and sodium compounds.

Certain kinds of modern soaps are made with animal grease. Others are made with oil from coconuts or cotton seeds or with other vegetable oils.

Scouring Materials

Soap, by itself, even if used as hot suds, does not remove the stains from steel knives and other metal things used in the kitchen. Such objects need to be scoured.

If you have dug dandelions with a stained steel knife, you have doubtless noticed how bright and shiny the blade became as you thrust it into the earth and pulled it out.

Fine sand, indeed, is one of the materials used in some kitchens for keeping knives and skillets smooth and bright. A finer powder may be rubbed from scouring bricks and used in the same way. A scouring brick may be a cake formed from a mineral like *feldspar*, or it may be a block of special baked clay. Such powder is good to use when we wish to avoid scratching the metal. Scouring powders are put into some kinds of soap during their manufacture.

You may recall reading in *Surprises* about plants called horsetails. Another name for such a plant is *Equisetum*. (*Equus* is a Latin word meaning *horse*, and *seta* is a

Latin word meaning *bristle*.) The plants
were given these names because they look like
whorls of stiff bristles. Some of the horsetails are also called scouring rushes. They

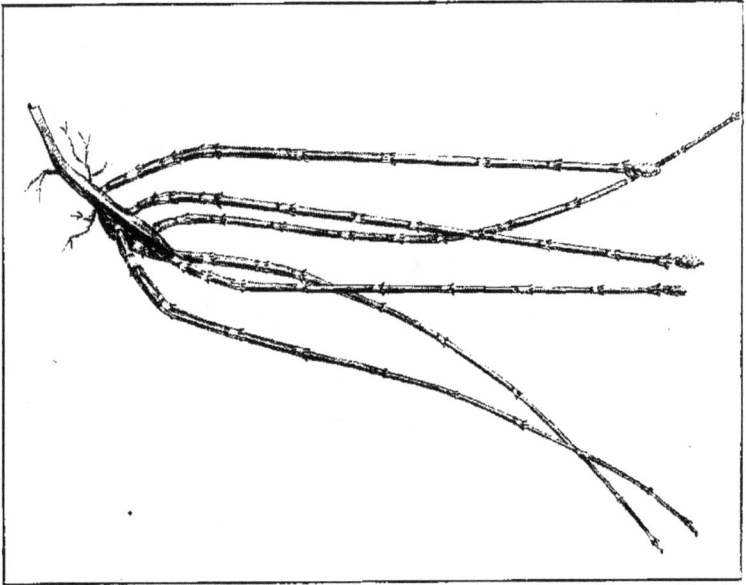

Scouring rush

have a flintlike material (silicon) forming a
frail skeleton throughout their mostly hollow
stems. Scouring rushes were formerly used
to polish wood and to scour metal objects.

Sometimes food burns in a kettle or pan and sticks to the bottom. Often scouring powder will not remove the burned stuff from the metal dish. Different sorts of scrapers can be used for this purpose. A tool of little metal rings linked together in a square is one such scraper. Or masses of rather fine steel fibers, called steel wool or metal sponges, are good for such work.

Brooms and Brushes

The materials and implements needed for keeping floors clean are as interesting as those used for cleaning dishes.

Brooms, or brushes with long handles, were given this name because they used generally to be made from broom plants.

Common broom (also known as *Scotch broom*) is a shrub with many stiff, wiry branches. It belongs to the same plant family as do peas and beans. Its pretty bright yellow flowers resemble pea blossoms in shape, and its seeds grow in flat pods with hairy edges.

Common, or Scotch, broom
The leaves of this plant are small.

Bushes of this sort are at home on dry, hilly places in the British Isles, in other places in Europe, and in parts of northern Asia. Years ago people brought some of the bushes to North America, where the plant thrives so well that it now grows wild in many sandy, barren places.

There are several attractive kinds of broom that people often have in their gardens. The blossoms of some of these are yellow, like those of common broom, while others have white flowers.

The tough, slender broom branches evenly trimmed and fastened together at the end of a handle make a very good brush for sweeping. Such brooms have been in use for hundreds of years. In this country and in many other countries, however, most ordinary brooms are now made of the seed heads of *broom corn*. This plant belongs to the Grass Family and is one kind of sorghum.

The stalks and leaves of broom corn look much like those of ordinary corn (Indian corn or maize). The seed heads are called

Seed heads, or brush, of broom corn
Does this plant grow in your state?

"brush" on account of their shape. Broom corn has been grown in forty or more states in this country. The crop is most successful, however, in those localities where there is plenty of rain during the early part of the growing season, but where there is usually dry weather at harvest time. In most places the crop is gathered before the seeds are ripe and when the brush is of a green color.

Various fibers other than those of broom corn are in common use for certain kinds of household brushes and brooms. Many brushes, for instance, are made of the coarse, stiff hair from horses' tails.

Even if you know how to sweep very carefully with an ordinary dry broom, you have doubtless noticed that some fine dust remains on a bare floor after sweeping. You may know, too, how well a wet cotton mop collects such dust; and perhaps you know why cedar oil is used instead of water on mops with which waxed floors are wiped or on cloths used to remove dust from wooden furniture.

Suction, or Vacuum, Cleaners

Have you ever used a glass tube, or pipette, with a rubber bulb, or cap, at one end and a small opening at the other (called a dropper or dropping tube or medicine dropper)? When you pinch the rubber cap between your thumb and forefinger, you press out some of the air that was contained in the tube and cap. If you then place the open end of the dropper in a glass of water and let the rubber cap expand, water rushes into the dropper. Do you understand what forces the water into the dropper?

Air, like other materials, has weight, as you have learned. It presses against everything it touches. Air on the surface of the earth at sea level is heavy enough to have a pressure of nearly fifteen pounds to a square inch.

The air inside the dropper presses as much as does the air outside while the dropper is full of air. But when you force some of the air out of the dropper by pinching the cap, there is not so much air pressure inside as there is outside. So the weight of the air on the water in the glass presses some of the water into the tube.

The water rises in the tube of the dropper as the water rises in the pipe of a suction pump, about which you read in another chapter in this book. You have, indeed, made a partial vacuum in the dropper by removing part of the air it contained. You may recall that there was also a partial vacuum in the pipe of the suction pump we have just mentioned.

Other substances besides liquids can be forced into a vacuum by air pressure. Of course the air all around you is pressing against everything else as much as it presses on the water in the glass or the water in the well. Men make the weight of air, or air pressure, do a great deal of work for them by using partial vacuums in different machines. A vacuum cleaner is one such machine.

Suppose that you have a little heap of chalk dust or other fine dust into which you place the open end of your dropper while you hold the cap tightly pinched together! What happens when you let the rubber cap expand? Air from outside the tube rushes into the partial vacuum inside and carries (presses) some of the chalk dust or other dust with it. Some of the dust is thus lifted from the heap into the tube.

You might call a vacuum sweeper or other vacuum cleaner a "dust pump" because a partial vacuum is produced inside the machine

by pump action or in some other way. If
dust or other loose litter is near the opening

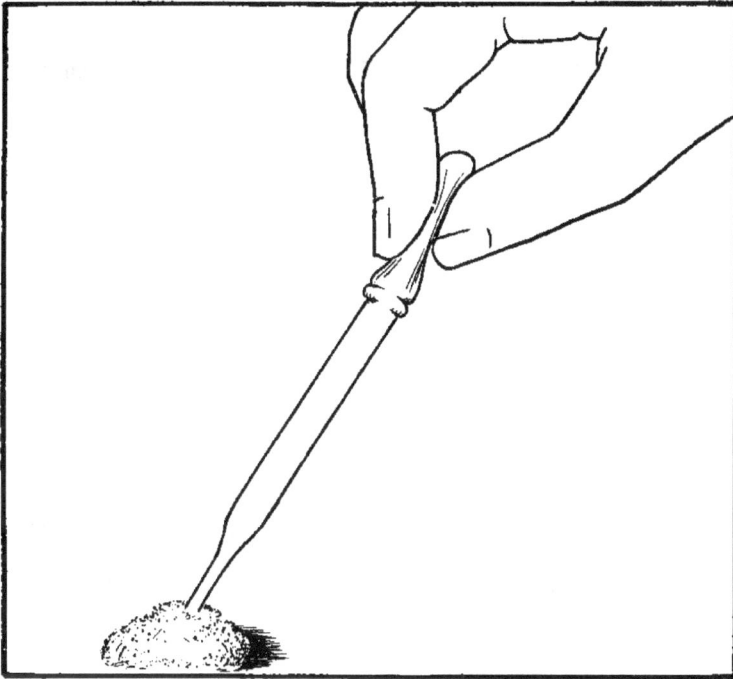

Air pressure forces fine dust into the partial vacuum
of the tube.

leading to the vacuum chamber, it is carried
into the chamber by the pressure of the air
that rushes in. This dust goes into a firm
cloth bag and is held there, while the air that

rushes into the sweeper can escape through the cloth.

For many purposes there is less hard work in using a vacuum cleaner than a broom or other ordinary brush. Since it removes so much of the dust, there is less in the air; so the air in the room is more healthful to breathe.

Of course many other cleaning materials and devices might have been mentioned in this chapter. But we think you may like to learn about some of the others for yourselves. You might, for instance, study the different methods of cleaning clothes and find the subject quite as interesting as any we have chosen for you to read.

Questions and Activities

Read on pages 271–272 of *Through Four Seasons** the directions in "Growing Some Mold." If you

* See page 424 for book list.

have never grown any mold, do so now in the way suggested on those pages.

In this chapter you read that there is silicon in the stems of plants called horsetails. What did you read about silicon in another chapter in this book? If you do not remember, look for this word in the index. Then find the page where the word *silicon* occurs and read again what is said about this substance.

Read as much of the book *Soap Bubbles* as you may find interesting while you are studying about soap.

Would you like to make some soap? This is not hard to do. You should not try to make soap, however, unless there is some grown person who is willing to show you how. You might hurt your hands with the potassium compound. The potassium compound can be bought, or it can be obtained by soaking wood ashes in water. The potassium compound used in making soap is *caustic* (capable of "burning" or destroying) and must not be allowed to touch flesh. It is sometimes called *caustic potash* or *caustic lye*.

When you suck lemonade through a straw, do you do all the work or does the pressure of the air on the lemonade in the glass help push the lemonade into the straw? Before you insert the tip of the straw into your mouth, there is air in the upper part of the straw (in that part that reaches above the lemonade). As you remove this air by sucking it out, do you make a partial vacuum in the straw? Do you think your mouth and the straw are a simple sort of suction, or vacuum, pump?

If it is convenient for you to get a dropper such as is mentioned on page 258, see if you can lift some chalk dust or other fine dust into the tube after you have removed some of the air by pressing on the rubber cap. Do you think you could call this dropper a very simple kind of suction, or vacuum, cleaner?

MATERIALS FOR CLOTHES

PART FIVE

Bud and blossom of cotton plant

COTTON

On another page in this book we remarked that the things we use every day are as interesting as exhibits in a museum. Perhaps you may like to think of a clothes closet as one room in the home museum. You may find it pleasant to learn about the plants and animals from which our clothes are obtained. There are some astonishing facts about the most common materials we wear. Shall we begin with the subject of cotton?

Cotton Trees and Shrubs

Cotton trees, as you may know, grow in tropical countries. If you were to visit certain warm places in Central and South America, you might see wild cotton trees with trunks as large as those of peach trees and with branches fifteen or twenty feet long.

People from Europe, however, did not see cotton trees for the first time in any tropical part of America. Before they knew there was such a continent as America, some Europeans traveled to India. They saw cotton growing there and called it "tree wool." The people of India had been making cotton cloth many hundreds of years before any European ever saw a cotton tree.

There are several different kinds of wild cotton plants. Some do not grow to be trees but live from year to year in the form of woody shrubs.

When seeds from cotton trees or shrubs are planted in countries not so warm as tropical places, however, the plants do not live from year to year. They do not become old and woody. They grow vigorously for one season, blossom, and ripen their seeds. Then they are killed by frosts. In such places cotton seeds must be planted each year.

The Cotton Belt

Cotton plants need about seven months of favorable weather in order to grow old enough each year to blossom and have ripe seeds. There are many places in the world where the climate is sufficiently warm for this growth.

The climate and soil of much of the southern part of the United States are suitable for growing cotton. Cotton is the most important crop in large areas of our southeastern states. Indeed, about half of the cotton of the whole world is grown there. (There have been times when two thirds of the cotton was grown in that region.) That part of our country is known as the *American Cotton Belt*.

Cotton plants in this cotton belt are not native to this country. The seeds were first brought from other countries and introduced here.

Several Kinds of Cotton

There are many varieties of cultivated cotton, but they all belong to a few main kinds.

Upland cotton is grown in the Cotton Belt more than any other kind. It thrives in inland places (away from the sea). Its flowers are white or cream colored. There are two sorts of fibers on the seeds. The longer fibers are from a little less than an inch to a little more than an inch in length. They are called *lint*. After these fibers have been removed by a machine called a gin (page 276), a short fuzz still clings to the seeds. These fuzzy fibers are called *linters.*

Sea Island cotton thrives in parts of Florida, Georgia, and South Carolina. The fibers are longer than those of upland cotton. Some seeds have fibers more than two inches long. When Sea Island cotton is ginned (put through a gin), the seeds are

Upland cotton plant

nearly bare, as there is little or no short fuzz on these seeds.

Sea Island cotton plants are taller than upland cotton, and their leaves are smoother. Their flowers are bright yellow when they open. Later they turn reddish.

This sort of cotton is used for making fine cloth and laces, as the fibers are soft and silky. They are strong, too; so they are used for the best grades of automobile tires.

Egyptian cotton is much like Sea Island cotton. It is not grown in the same places because it is not healthy in the southeastern states. It is likely to become sick with a disease called black arm. It has been grown successfully, however, in Arizona.

The several kinds of cotton that have been mentioned have all been brought to this country. They have been grown only since white people settled in the South and started cotton plantations.

There is reason to believe, however, that Indians grew one kind of cotton in the Gila Valley, Arizona, two thousand years or more before white people came to America. Evidence of this has been found in ancient ruins. Pima Indians, living in that part of Arizona today, still grow a cotton that is thought to be the same kind. Hopi Indians grow it, too, and it is called *Hopi cotton*. This sort of cotton has small lemon-yellow flowers with black dots. The lint is white, strong, fine, and silky.

Cotton Fibers

You are familiar with the seeds of certain plants having fluffy fibers for sails that enable the seeds to float through the air. They are carried by the winds when they are ripe and ready for their journey to places where they can settle and grow.

As you may know, the sailing seeds of

aster and goldenrod and dandelion and many other plants do not grow in pods. Others, like those of the milkweed and cotton, are held inside closed pods until they are ripe, when the pods open and set free the seeds.

Cotton pods are called *bolls*. The word *boll* is really the same word as *bowl*, but it is spelled in an old-fashioned way. A boll, or a bowl, is a round vessel. So this is a good name for a round seed vessel as well as for a round dish.

The seeds of cotton plants have many fine fibers, by means of which they can be carried long distances by the wind. This scatters the seeds far and wide, so that cotton plants may have a chance to grow in new places. Of course this is a great advantage to wild cotton plants. Their seeds are very useful in spreading cotton through favorable tropical regions.

Perhaps it was many thousands of years ago that people living in some warm countries

learned that they, as well as the plants, had
a use for the fibers on cotton seeds. They
found that they could fashion the fibers into

Courtesy U. S. Dept. of Agriculture
Closed and open cotton bolls

thread or yarn with their fingers and could
weave the threads into cloth.

A cotton fiber is not quite even and
straight. It has a twisted shape. You need
to look at a fiber through a magnifying glass
to see the twists. When cotton fibers are

twirled together, the twists help hold the parts of the thread and make it strong.

Cotton Gin

For many centuries all the work of pulling the fibers off the seeds had to be done by hand. This was a slow task. Even after people began to grow cotton in our southern states, they did not plant much in the early days. It took so long to separate the fibers that the best they could do was to send just a few bales to factories in England each year. In those days cotton cloth was very expensive.

Eli Whitney was the man who changed this state of affairs. He invented the *cotton gin* in 1792. This was a machine which could be used to separate the seeds from the fibers very rapidly.

After that people planted more and more cotton in the South. The ripened seeds are ginned, and the fibers are sent to mills in huge bales. (The seeds themselves are not wasted, and you may know some uses that are made of the oil they contain.)

Reproduced by permission of the Philadelphia Commercial Museum

An early cotton gin

The Cotton Trade

There came a time when the United States produced much more cotton than any other country. Much of the baled cotton is sent to mills in this country, where it is made into cloth. Vast quantities of the baled cotton, too, are sent to other countries.

Of course it is pleasant for the people in this country to sell so much cotton and have the money to use. This crop has done much to help make the United States a prosperous country. But the people in certain other countries think it would be better for them if they could grow all or part of their own cotton, instead of buying so much from us. So in recent years more cotton has been planted in China, in the Sudan north of the equator in Africa, in Australia, and in some other tropical or warm temperate regions.

Diseases of Cotton Plants

It would be easier to grow cotton if the plants were always well. They are so likely to become sick, however, that it is often necessary for growers to take great care to prevent or control certain diseases.

Black arm has already been mentioned on a previous page in connection with Egyptian cotton. This disease is caused by bacteria. (You may have heard of blackleg, a bacterial disease of potato plants.)

Wilt is a disease which causes heavy losses to the cotton crop in the United States. It is troublesome in the sandy soils of the Cotton Belt. There have been years when about nine tenths of the plants in some of these fields wilted and died. Wilt is caused by a fungus. The fungus enters the roots. It grows in the water-carrying vessels of the plant. It plugs these vessels and shuts off the water supply. So the plant wilts.

Root knot is a disease causing the roots of a sick plant to be covered with galls. These galls vary from the size of a pinhead to a thickness of more than half an inch.

Young galls are whitish, but later they turn brown and decay. They cut off the food and water supply of the plant.

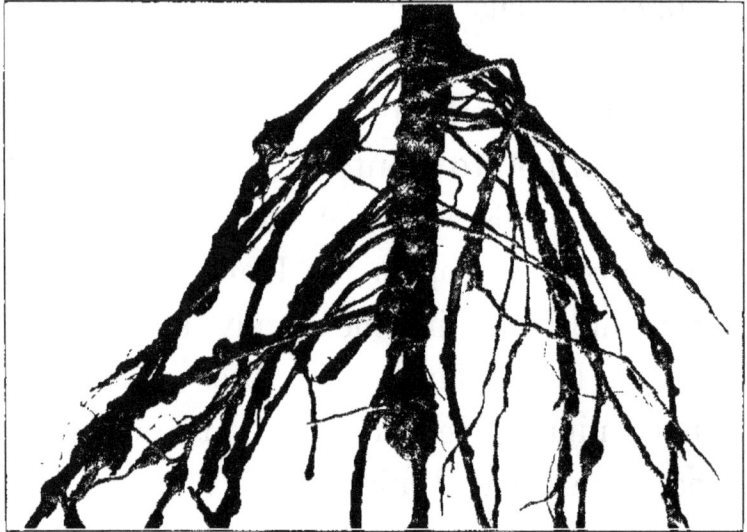

Courtesy U. S. Dept. of Agriculture

Root knot on cotton

Cotton is not the only kind of plant to have root knot. Cucumbers, potatoes, some kinds of peas, and many other plants may have this same disease. It is caused by *eel-worms*. These tiny worms bore into the roots and irritate them. Galls grow in the injured places. If you break open one of these galls, you may be able to see some of the eelworms. They are so small, however,

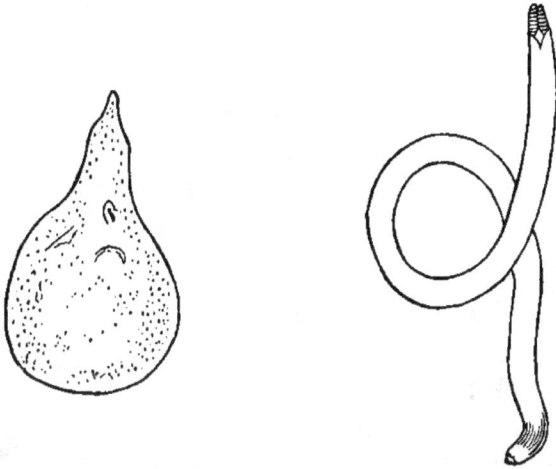

The way eelworms look when seen through a magnifying glass

that you need a magnifying glass to see them plainly.

Insects That Trouble Growers of Cotton

Diseases are not the only troubles that cotton growers must meet. Destructive insects bother them, too. Aphids, beetles, caterpillars, and others are to blame for certain sorts of damage.

The *bollworm* has destroyed so much cotton that it has been regarded as one of the most important insect pests of the world. It is the caterpillar, or larva, of a moth. It

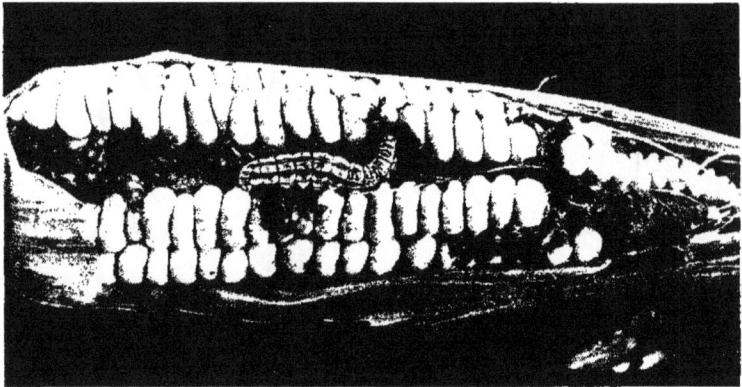

Courtesy U. S. Dept. of Agriculture

A bollworm eating corn on the cob

eats the flower buds and young bolls of the cotton. This same insect likes other food, too; it is known as the *corn ear-worm* and *tomato worm* because it chews its way into ears of green corn and helps itself to juicy tomatoes.

It would be easy to fill a book with accounts of different insects that have troubled cotton growers. But after the boll

A cotton boll weevil on a cotton boll

weevil arrived, these people were so worried about the new pest that they seemed to forget about most of the others.

The *cotton boll weevil* is a beetle that came from Mexico to Texas. A beetle of this sort nibbles a hole in the green boll with its beak and then pushes eggs into the hole. The young grubs which hatch from the eggs eat the contents of the boll.

This unwelcome intruder spread so rapidly and did so much damage that people in Texas, Louisiana, Mississippi, and Alabama were discouraged. Many cotton pickers had no cotton to pick, wealthy growers became poor, banks failed, and everybody who had any interest in the affairs of the Cotton Belt was worried.

In spite of all the worry, however, there were some men who had courage to try to beat the boll weevil and have the cotton crop for themselves. Texas offered a prize of $50,000 to anyone who could discover a satisfactory way to control the pest. State and United States Government scientists studied to see if there was any way to make cotton a profitable crop in spite of the weevil.

Energetic growers tested the most sensible methods as soon as they knew about them. In time people learned that, if they planted

Courtesy U. S. Dept. of Agriculture

Dusting a cotton field from an airplane

poor cotton lands to other crops and took greater care of the cotton on the better land, they could still get cotton to harvest.

But, in order to do this, the plants must be dusted with a poison powder to kill the

Courtesy J. M. Robinson

"In profound appreciation of the boll weevil and what it has done as the herald of prosperity, this monument is erected by the citizens of Enterprise, Coffee County, Alabama."

weevils. The first of these dusters were pulled by horses. Later airplane dusters were used.

Indeed, the fear of the boll weevil spurred growers to make so many improvements in cultivating cotton that old slipshod ways were discarded. Newer and better methods were used. And more kinds of crops were planted.

People in the very states that had been most troubled finally said that, after all, it had been a good thing that the weevil had arrived and frightened them into improving their farms. They felt like saying, "Thank you," to the very insect they had looked upon only as an enemy. That is how it happened that a beautiful statue was erected to the boll weevil.

Helpful Insects

The insects mentioned so far in this chapter are those that injure cotton plants. We should not forget that there are other insects that help these plants.

If you should visit a cotton field in blossom time, you would see winged guests flying from flower to flower. You would hear the humming of their busy wings. You could see that they were there for nectar or pollen.

And perhaps you would understand that they carried pollen about on their dusty little bodies. In this way pollen is taken from a stamen of one flower to a pistil of another. This is most important, for it enables the beginning seeds in the second flower to develop.

Thus we have depending on the visits of pollen-bearing insects: the lives of young cotton plants; wages for cotton pickers; wages for men working at cotton gins and in cotton mills; profits for cotton growers and mill owners; and cotton garments for you.

Of course the insects that visit cotton flowers to drink nectar or eat pollen know nothing about the dusty gifts they carry from plant to plant. But it is not fair that people should forget the debt they owe to these volunteer laborers in the cotton field. It is true that some insects cause great damage to this important crop. On the other hand, there would be no cotton crop at all if it were not for such insects as take live pollen to the forming seeds.

Questions and Activities

Why does the same kind of cotton plant live for years and blossom many times in tropical countries when it may blossom but once in a warm temperate region like our southern states?

Make a list of all the words in this chapter that are printed in *italics*. Use any ten of these words or expressions in sentences of your own.

Read the chapter called "Healthy Potatoes" in the book *Through Four Seasons*.* Notice

* See page 424 for book list.

especially the parts about bacteria and fungi. Name one potato disease caused by a kind of bacterium. Name a potato disease caused by a fungus.

Let each member of the class choose one of the references in the following books:

(1) "Cotton," pages 147–152 in *Introduction to World Geography*

(2) "Seeds That Float with Filmy Sails," pages 49–53 in *First Lessons in Nature Study*

(3) "The Cotton Plant and Some of Its Relatives," Chapter VI in *First Lessons in Nature Study*. (Notice the account of cottonseed oil given on pages 123–124 of this chapter.)

(4) Any of *Cotton and Other Useful Fibers* that interests you. Also look at the pictures on the pages about cotton.

After you have read one of these four references, tell the members of your class what you have learned. If it is convenient for you to do so, show your classmates the pictures in the book that concern cotton.

LINEN

Pure Irish Linen

It is said that in one year alone enough linen cloth was woven in Ireland to carpet a path three feet wide around the earth. Labels with the words "Pure Irish Linen" have been placed on vast quantities of this kind of cloth. You can see such labels on handkerchiefs and other linen articles on the counters in stores.

Ireland is, indeed, famous for the products of its linen mills. But, after all, this fame is rather recent. For thousands of years before any Irishman ever saw a linen thread, the people in India and Egypt were making good linen cloth. Of course that was long before the days of printed labels.

Photo by L. M. A. Roy, from R. I. Nesmith and Associates

Spinning flax

Common Flax

Linen, as you know, is made with fibers from the stems of flax. Flax is not a native plant in Ireland. Seeds were carried there by men who made voyages in sailboats many years ago. It is thought that these old-time traders saw flax plants and linen cloth in Egypt and took some of the smooth, glossy brown seeds from that country to Ireland.

Of course you know that you do not need to go as far as Ireland or Egypt now to see these plants, for in time the hard, shiny little seeds of common flax were carried across all the seas and planted on all the continents except Antarctica as well as on many islands.

It was a common practice in the early days in this country for farmers to grow some flax on their own land. Perhaps you have seen one of the old spinning wheels on which the women of those days spun linen yarn to use when they wished to weave cloth.

In time some of the eastern farmers began to move west. They found that flax was good to grow for their first crop after the prairie sod was plowed. As men have moved westward farther and farther, many miles of prairie land have been plowed and there have been great fields of flax to sow and reap.

You might easily walk by some of the slender-stemmed flax plants with their small, narrow leaves without noticing them at all. That is, you might — if they were not in blossom. You may have heard some one say, "I remember, when I traveled through South Dakota on a train, that we passed fields blue with the flowers of flax." Who, indeed, could forget so beautiful a sight?

Very little of all this northern and western flax, however, is grown for fiber. In this country most of the flax is grown for seeds. The seed of the flax plant, like that of cotton, has a valuable oil in it, but it is put to different uses. You read in another chapter about linseed oil.

When flax is grown for the seed crop, it can be cut with the same sort of machines that are used in grain fields. Later the seeds can be threshed out with ordinary threshing machines.

Fiber Flax and Seed Flax

In some countries where flax is grown for fibers the seeds are also saved. In such places the seeds are considered a by-product, because the fibers are of more importance. Fiber flax is planted thickly, and there are very few flowering branches to have seeds, Tall varieties are good to grow for fiber. The best fiber crop is harvested before the latest seeds are ripe.

In the great northern and western flax fields in the United States, it has been the usual practice, as you know, to plant a variety of flax that yields a large crop of seeds. The fiber in the stems of short varieties of flax, which are raised for seed, is not a good grade.

Recently farmers in Oregon have been interested in a flax-fiber industry. There flax is being grown as a fiber crop, though the flaxseed is saved for oil, as is done in other countries.

There is no reason, so far as climate is concerned, why excellent fiber flax should not be grown in this country. Common flax does not need six or seven frostless months, as cotton does. It will thrive in localities where there are about three and one half months of good growing weather. During that time the plants attain their growth, blossom, and have ripe seeds. Indeed, flax flourishes best in places where the climate is not too hot. It would not grow so well in the Cotton Belt.

Bacteria Help in the Retting Process

You have learned of various ways in which bacteria are useful to people, but you may not know how they serve the linen industry.

The pulled flax stems are placed in a wet field or soaked in a pool or slow stream while they rot, or ret. During this process the fibers become loosened from the outer layer, or bark, of the stem. The water itself cannot soften the stems enough for this change to take place. Great numbers of little bacteria grow in the wet flax. Their presence is necessary to rot the parts of the plant that hold the fibers together.

Rotten flax has an unpleasant odor. You may be interested to know that some scientists are experimenting to see if they can find satisfactory methods of softening the stems without the help of bacteria. They hope that certain chemicals may be used that will loosen the fibers. If the stems can be softened properly without permitting them to rot, the manufacture of linen will be a more agreeable task.

After the stems have been retted, the fibers are separated from the other parts of the stem by machinery. The fibers may be from one to three feet long, depending on the height of the plants and the care in handling the stems.

The best fibers are used for cloth. The coarser, poorer fibers are made into rope and string or are used for stuffing furniture and other articles that need to be padded.

Flax Wilt

You have heard of cotton wilt. You may be interested to know that flax also may become sick with wilt. Flax wilt is caused by fungus, but this is a different species from the fungus that causes cotton wilt.

Flax suffers worst from this disease when a flax crop is grown for a number of years on the same land. This is because the flax-wilt fungus lives in the soil for a time and is ready to attack flax if it grows in this soil too often. Such soil is said to be *flax sick*.

For this reason, it is a good practice to give the land a rest from flax and grow other crops on it part of the time. It is often better to plant other crops for four or five years on ground after flax has grown on it, before it is used again for flax.

Flax Seeds from South America

Botanists and plant doctors have studied different varieties of common flax to see whether they can find varieties that remain well even if grown on flax-sick soil.

There is a difference in flax plants. Some will thrive in the same soil where others wilt and die. Such strong plants are said to be *resistant* to wilt.

Scientists who wished to experiment have secured seeds from different countries for

Counting out two hundred seeds from each sample of Argentina flax

testing. A United States botanist who is very much interested in flax visited Argentina to study flax fields in that country.

One purpose of his trip was to get samples of
seeds suitable to test.   Scientists in Argentina

Courtesy H. L. Bolley, North Dakota Agricultural Experiment Station

**Flax samples from Argentina on flax-sick soil in
North Dakota**

The plants in some rows have died.   The more resistant
plants are alive.

gave the visiting botanist more than four
hundred samples of seeds.

Later many of these seeds were tested in a flax garden at the North Dakota Agricultural College. Two hundred seeds of each sample were planted on flax-sick soil. Certain samples were more resistant than others. Some were strong and sturdy while others died.

QUESTIONS AND ACTIVITIES

Linum

As you already know, the Latin name for a flax plant is *linum*. An English form of this word is *line*. (Fishlines and other lines and ropes have been made with flax fibers.) Notice that the first three letters of the Latin name (*lin*) are used in different English words.

(1) Why is the word *linnet* a good name for a bird that feeds on flax seeds?

(2) Explain the word *linseed*. (Use a dictionary or an encyclopedia for help if you wish.)

(3) Explain the word *linen*.

(4) Look for the word *linoleum* in a dictionary or an encyclopedia. See if this kind of floor covering has any connection with flax.

Red, White, and Blue Flax

Common flax is beautiful as well as useful. People often have this plant in their flower gardens for the sake of the blue blossoms. But there is a species of flax with red blossoms and one with white blossoms that are also grown in flower gardens. Ask if you may have red, white, and blue flax in your school garden.

Some Other Plant Fibers

Choose to learn a little about materials made from one or more of these plants: hemp, jute, sisal, Manila hemp. You can find these words in the index of *Cotton and Other Useful Fibers** and of *Introduction to World Geography.* You may look for them in other books about fibers and in other geographies if you like. Write about the plant you choose to study. Name a country in which it grows. Tell how its fibers are used.

Read the chapter "Flax and Some Other Fiber Plants" in *First Lessons in Nature Study* and notice the pictures on pages 39, 41, and 43 in *Surprises.*

* See page 424 for book list.

SILK

Spider Silk

Soft and beautiful cloth is woven from silk spun by certain spiders called by the rather pretty name of *Nephila*.

If you travel sometime to the Indian Ocean and visit Madagascar, perhaps you may see for yourself how the silk makers there take care of their spiders. You may have read about these Nephila spiders, which are kept in tiny stalls that hold them in place without hurting them. While a number of spiders are thus held in position close together, the silk fibers that they spin can be twisted into a thread that is thick enough and strong enough to wind.

After the spiders have been spinning for a

while, they are given a vacation in a garden. They are permitted to feed and rest for some time before they are again placed in their little spinning stalls.

Cloth of spider silk is so soft and fine and strong that people in different countries have tried to see if it would pay them to manufacture articles of this substance. This matter has been studied in our country as well as in China and France.

Nephila spiders thrive chiefly in tropical and semitropical countries. They cannot be kept long in captivity, and it is not always easy to supply them with proper food.

There are, indeed, many difficulties in taking care of Nephila spiders and handling their fibers. For this reason no spider-silk industry has really been established except in Madagascar. So articles made of this material are rare and costly.

Silk Glands and Spinnerets of Spiders

Of course you know that it is not necessary to go to Madagascar to see some kinds of spider silk. You can find a spider web any summer day if you hunt in places of the right sort. You would not need to travel a mile to see one.

Although fibers made by Nephila spiders are best for weaving into cloth, other kinds of spiders make silk that serves their own purposes. As you may know, they make silken sacs for their eggs, some spin webs for insect traps, and some weave tunnels for shelter.

This silk is produced inside a spider's body in silk glands. It is a sticky liquid while it is in the glands. From there it passes to the spinning organs. The spinning organs, or *spinnerets*, of spiders are too complicated to be described fully in this chapter; but you will wish to know a little about them. The spinnerets are small, jointed organs near the tip of the spider's abdomen. Six is the number that most spiders have. They sometimes look like a tuft of little tails on a spider's body.

This spider has made a silken sac to hold her eggs.

The sides of a spinneret are hard, but there is a soft, thin membrane at its tip. The fluid silk from the spinning tubes passes through minute openings in this membrane. When the fluid touches the air, it at once hardens and changes into a fiber.   It is then ready for the spider to use.

Caterpillar Silk

Fortunately for people, spiders are not the only animals that spin silk.   The larvae (young) of many insects produce and use silk in various ways.

Caterpillars, the larvae of moths and butterflies, are famous spinners.   Some species produce a fiber only a few inches long, while others may spin a fiber which, if stretched in a straight line, would reach nearly half a mile.

There is one family of moths the larvae of which are called *giant silkworms*.   You may know about the *Luna* and *Cecropia* moths, two of the large and handsome moths that belong to this family.   The picture on page 309 shows that of a third member of this family, known as the *Promethea* moth.

Silken cocoon made by one kind of giant silkworm
and the moth that came out of the cocoon

A giant silkworm spins a very good cocoon of silk when it is a full-grown caterpillar and ready to change to the next stage in its life. After the cocoon is made, the caterpillar becomes a pupa, which rests inside the cocoon. In time the pupa develops into a moth, and then the insect makes its escape through one end of the cocoon.

Wild Silk

Good coarse silk cloth is woven from silk fibers spun by giant silkworms. Some species of this family live in this country. Others may be found in Europe and Africa. It is in Asia, however, that the cocoons of giant silkworms are valued as a source of silk.

People in northern China collect the cocoons of giant silkworms that live there. The silk taken from these cocoons is called *wild silk* because for many years the people took no care of such caterpillars. They merely collected cocoons that "wild" caterpillars made.

After a time the Chinese began to rear giant silkworms, caring for them and feeding them until they made their cocoons. So now these caterpillars may be called "domesticated" or "tame" instead of "wild." One kind lives on oak leaves and another on the leaves of ailanthus trees.

The kind that feeds on ailanthus leaves is named *Cynthia*. Long ago eggs of the Cynthia moth were taken from Asia to Europe so that these caterpillars might be reared there. In 1861 this species was brought into this country for the same reason.

Neither Americans nor Europeans have made a business success with giant silkworms though more than a million dollars' worth of this sort of silk has been exported from northern China in a single year.

Cloth that is woven from silk spun by giant silkworms is called by the different names of *pongee, rajah, shantung,* and *tussah.* You may be interested to see some of these silks in stores.

The Mulberry Silkworm

Most of the real silk cloth in the world is made from fiber that is spun by one kind of tame caterpillar. This is called the Chinese silkworm or the mulberry silkworm or just the silkworm.

The silkworm's skin is smooth and yellowish white. This caterpillar likes the leaves of a mulberry tree better than any other food. When it has eaten as many mulberry leaves as it can, it is about two inches long. It is then ready to spin its cocoon and change to a pupa. The cocoon is yellow or white, and it is made without any doorway.

Silkworms, moths, and cocoons

The insect lives as a pupa inside the cocoon for about three weeks (or more if it is kept in a cool place). By the end of that time it has its wings and is ready to come out of the cocoon.

Although the cocoon has no doorway, the moth can get out. It squirts some liquid against one end of the cocoon. This liquid softens the gum that holds the silk together. It is then easy for the moth to push its way through the wet end of the cocoon. The moth is cream colored, with some faint brown lines on the forewings.

There are interesting old tales about silkworms in China. One story is that the Empress Si-Ling-Chi was the first to rear the caterpillars and reel silk from the cocoons. She was given great honor and called Goddess of the Silkworm.

Centuries ago Chinese people found that, if they put the cocoons of the mulberry caterpillars into hot water, the gummy stuff on the silk would soften and the fiber could be unwound. They learned that fibers from several of these cocoons could be twisted together into thread and then woven into cloth.

This way of getting silk fiber to use was long kept a secret by the Chinese. They practiced weaving and dyeing and embroidering until they could make silk cloth with pictures of flowers and dragons and people on it.

The silk robes that the Chinese princes wore were the most beautiful clothes in all the world. The Chinese sold some of their silk cloth to travelers from other countries, who paid great prices for it. This cloth was so famous that China was called the Land of Silk.

People from other countries could buy all the silk cloth they could afford, but they could not buy the thread or find out where it came from. That was a secret that the Chinese kept for many hundreds of years. No one in China dared to tell about it.

If people in other countries desired silk thread to use, they had to ravel it from silk cloth that came from China. Some of them thought that this thread was made from plant fibers, as are linen and cotton threads. Some thought that it was made from the fleece of sheep in a secret way.

In one way and another the secret of getting silk at last reached the people living in different countries. Mulberry trees were planted in many places where they had never grown before. Their leaves were taken off and fed to the tame silkworms.

When James I was King of England, he sent some mulberry trees and silkworm eggs to Virginia and told the people there to turn their attention to silk instead of tobacco. (He also told them to grow flax for linen.)

They did as they were directed for a while. But it proved to be cheaper to import silk to America than to produce it here.

If you read the accounts of different attempts to rear silkworms in America, you will find much to interest you. You will read that Benjamin Franklin wished to persuade the people in Pennsylvania to grow silkworms.

In many other states, too, people tried this business. For a while it was the fashion for American women to wear silk gowns made from silk they had unwound from the cocoons of silkworms they had reared themselves.

You may like to learn whether mulberry trees have ever been planted for silkworms in your state. In most parts of this country people gave up trying to rear silkworms some time ago. But, if you happen to live in California, you may know that a few years ago a man in that state planted many thousands of mulberry trees for silkworm food.

Although few people in the United States are now interested in rearing silkworms, more silk is woven into cloth in American mills than in those of any other country. The raw silk is sent to our mills from other countries (chiefly China, Japan, and Italy) after it has been unwound from the cocoons.

Silk Glands and Spinnerets of Caterpillars

As you may have read in the book *Surprises*,* there are two silk glands inside the body of a caterpillar. These connect into a narrow tube that opens through the caterpillar's lower lip. As you probably know, the silk that comes out of this opening is a sticky liquid while it is in the glands, but as soon as it touches the air it becomes stiffened into a thread.

* See page 424 for book list.

QUESTIONS AND ACTIVITIES

Read about spiders in Chapter VIII of *First Lessons in Nature Study*.* Then tell how the young of some spiders use silk fibers when they travel. Name some seeds that use their fibers in a similar way.

Hunt for a cocoon made by a giant silkworm if you live in the country or go to the country for a visit. If there is a pupa in the cocoon, keep it until the moth comes out. Put the empty cocoon into boiling water. Then see if you can unwind silk fibers from the cocoon and twist them into a thread.

Is there a silk mill near your school? If there is, ask your teacher if it is convenient for her to arrange to have your class visit this mill.

There are four chapters about caterpillars that spun silk in *Hexapod Stories*. There is one chapter in *Holiday Meadow* and one in *Holiday Hill*. Read one of these chapters and tell your classmates what the caterpillar you read about did with its silk. Did it make a cocoon or did it use its silk in some other way?

* See page 424 for book list.

Are you willing to take good care of some kind of caterpillar? If you are, keep one and feed it until it is ready to spin. (Give it the same kind of food it is eating when you find it.) Watch it spin and notice how it uses its silk.

RAYON

It is perfectly natural for a caterpillar to make silk. All it needs to do is to let a sticky liquid run out of its silk glands and pass through its spinneret. As soon as this substance reaches the air — there is a silken fiber all ready to use in making its cocoon.

Many men who have watched silkworms at this easy task have doubtless wished they could do as well as these caterpillars. It is not surprising that some of them should have studied the material these insects use with the hope that they might make something as good.

About two hundred years ago René de Réaumur, a French scientist, wrote a book about insects. In this book, he said: "Silk is only a liquid gum which has been dried. Could we not make silk ourselves with gums and resins? . . . It does not seem impos-

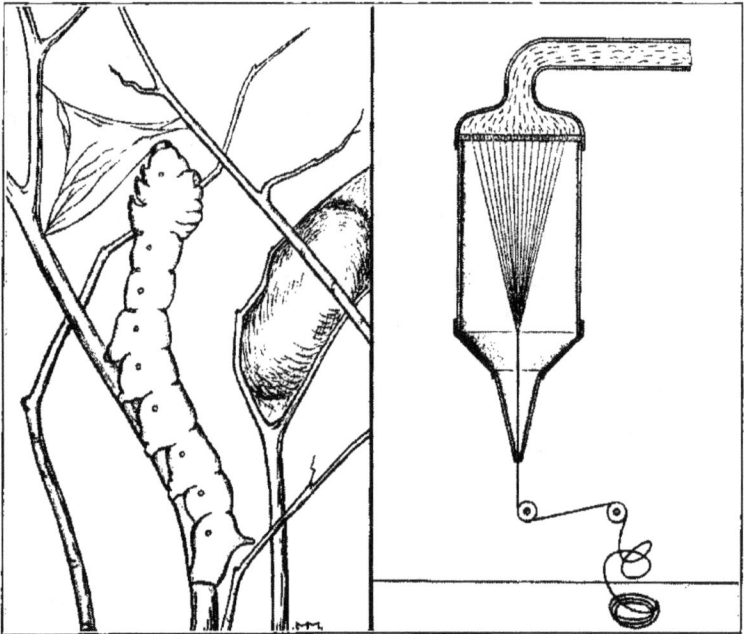

Natural silk and artificial silk coming out of spinnerets

sible to spin them into threads . . . when we consider to what extent art may be carried."

Cellulose

About seventy years ago so many silk-worms died of disease that people feared there might be no well silkworms left to make silk for them.

The situation was so serious that a French chemist, Louis Pasteur, devoted two or three years to the study of the dreaded silkworm sickness. He investigated the cause of the trouble and found ways to take care of these valuable insects so that they could be kept healthy.

While the famous French chemist and his assistants were studying silkworms, they became interested in other matters besides silk-worm diseases. They gave considerable attention to certain changes that went on inside the bodies of these caterpillars.

They examined the sticky liquid in the silk glands. They found two important substances in it, each made from *cellulose*.

One of these is the stuff called *fibroin*, from which the fiber itself is formed. The other is *sericin*, a waxlike material also found in the fiber.

Caterpillars take cellulose into their bodies when they eat leaves. There is a great deal of cellulose in plants. The bodies of plants are made up of tiny cells. The cell walls are composed of the substance known as cellulose.

Cellulose will not dissolve in water, as salt or sugar will. It is well for plants that this is so. It would be very unfortunate if the cell walls in all the leaves should "melt" whenever rain or dew touched them.

But cellulose can be changed into different compounds by chemical processes. The caterpillar's body may be regarded as a tiny chemical laboratory in which the cellulose becomes changed into silk-making substances. These are stored in the silk glands.

When the caterpillar is ready to use silk, it forces some of the silk-forming compounds through its spinneret. The air hardens them. Then the silk-forming compounds, fibroin and sericin, are no longer liquid. They are solid in the form of fiber ready for the caterpillar to use.

Men Make Fibers from Liquid Cellulose

After the chemists had studied the processes by which caterpillars make silk, they began certain experiments in their laboratories. They wished to see whether they, too, could produce useful fibers from cellulose. They poured different chemicals over cellulose to learn which would dissolve it. At length they were successful in dissolving cellulose so that it was a liquid.

The next step of course was to change the liquid into fibers. The experimenters made a little machine, called a spinning machine, with a spinneret equipped with tiny openings. They squirted the sticky liquid cellulose through those openings, expecting that the stuff would harden into fibers as it touched the air. It did.

Thus, after a great deal of study and trouble and expense, learned scientists succeeded in producing useful fibers similar to those a caterpillar can make as naturally as breathing.

Four Processes

One of the young men who studied under Louis Pasteur was Count Chardonnet. He became interested in cellulose and tried for many years to perfect a process for making chemical fiber (or chemical yarn as it is commonly called). His earliest experiments were made with mulberry leaves. At last he succeeded in finding a method that was good enough to patent. This was about 1884.

Chemical yarn had been made from cellulose before this date. Long before, indeed, a man in Switzerland had made from dissolved cellulose what he called artificial silk.

None of the earlier ways of making it was suitable for manufacturing the material on a large scale. Count Chardonnet's methods, however, were so satisfactory that, with some changes, they are still used. Chemical fiber, sometimes called artificial silk but now more generally known as *rayon*, is still made by what is called the *Chardonnet process.*

Later three other good methods of making chemical yarns were perfected. So at the time this book is being written there are four important processes for manufacturing yarns from dissolved cellulose. Much less is made by the Chardonnet process than by the three other methods.

These processes differ in the chemicals that are used and in some other ways. All four methods are very complicated. You would not be pleased if you were asked to learn about them now. Some years later, if you decide to be chemists, you may be interested to study the subject of chemical yarns. It is enough for you to remember at present that cellulose is dissolved in certain chemicals, and that the sticky liquid is squirted through spinnerets and hardened to fibers.

Hardening the Fibers

There are two ways of hardening the fibers as the liquid is forced through the spinnerets.

One way is to squirt the cellulose solution into air (which is what a caterpillar does). Then the liquid goes off in vapor in the air (evaporates) and leaves the solid fiber behind. This process is sometimes called "dry spinning."

The other way is to squirt the cellulose into a "bath" tank. There are chemicals in the tank that change the cellulose so that it leaves the "bath" as solid fibers. One name for this process is "wet spinning."

Cotton Linters and Wood Pulp

Although silkworms can live by eating leaves of a few other trees, they always take mulberry leaves for food if they can get them. Naturally, then, they use mulberry cellulose when they make silk. Similarly, other caterpillars use cellulose from any leaves they prefer.

Plants of course have cells in all parts of their bodies. So there is plenty of cellulose in the cell walls in their roots and stems and seeds, as well as in their leaves.

You will remember reading how flax stems are retted to obtain textile fibers (fibers for weaving). These fibers are almost pure cellulose. Linen yarn is made from them. Rayon, too, could be made from flax fibers if anyone wished to do so.

The fluffy fibers on cotton seeds are almost pure cellulose. They are suitable for rayon as well as for cotton yarn.

Have you noticed the broken fibers in blotting paper or filter paper? These fibers, too, are almost pure cellulose. They are taken from wood pulp. Such fibers are made into rayon as well as paper.

Although cellulose from almost any plant can be changed to rayon, cotton linters and wood pulp are about the only materials that are used for this purpose at the present time.

Cotton linters, as you will recall, are the short, fuzzy fibers left on certain kinds of cotton seeds after the long fibers are removed by ginning. Chemists have learned how to use linter cellulose in manufacturing many important things — rayon among them. Indeed, great quantities of linter rayon are manufactured each year.

Even more rayon, however, is made from wood pulp than from cotton linters. We shall now follow some of the steps in manufacturing this important man-made stuff from wood pulp.

Rayon from Spruce Logs

Spruce trees are cut in the forests of northern countries. The logs are hauled to river

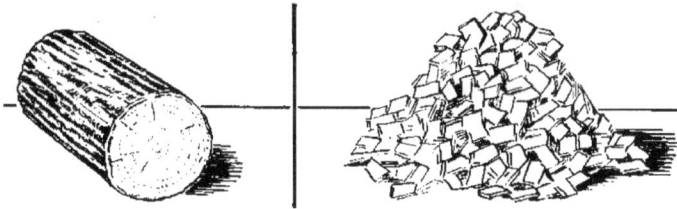

Spruce wood and spruce-wood chips

banks during the winter, where they are piled and left until the water is running in the spring. Then they are rolled into the streams and are carried by the running water to pulp mills.

After the logs arrive at a pulp mill, the bark is stripped from them. Then they are cut into small chips.

The chips are boiled in a chemical solution. This process removes from the chips the resin and the hard woody substance called

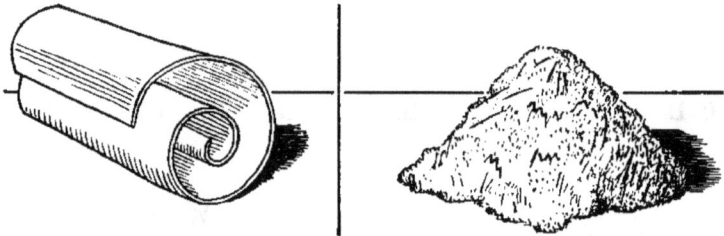

Wood pulp and wood-pulp crumbs

lignin. The cellulose is left in a pulpy sort of mass.

This wet, pulpy cellulose is bleached and washed and then pressed into sheets. They are white and much like thick sheets of blotting paper. They are called wood pulp, and the special name for the kind suitable for rayon is "bleached sulfite pulp."

The sheets of wood pulp are sent to the rayon factory. Here they are treated with caustic soda. Caustic soda is a chemical that was used in another way by an English chemist named John Mercer. He found

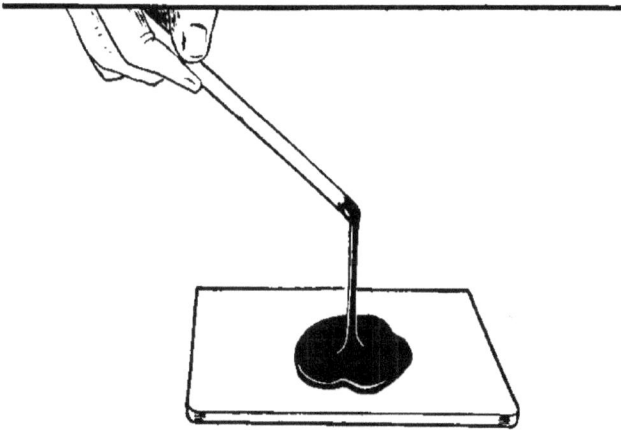

Sticky solution ready for the spinnerets

that, when cotton fibers are treated with a solution of caustic soda and then dried while being held tightly, they will have a sheen like that of silk floss. We call such cotton *mercerized* in honor of the man who made the discovery about 1840.

In making rayon from the cellulose from wood, a similar solution of caustic soda is used to change the cellulose. The changed cellulose is washed, drained, and then ground into fine, fluffy bits that are called *crumbs*.

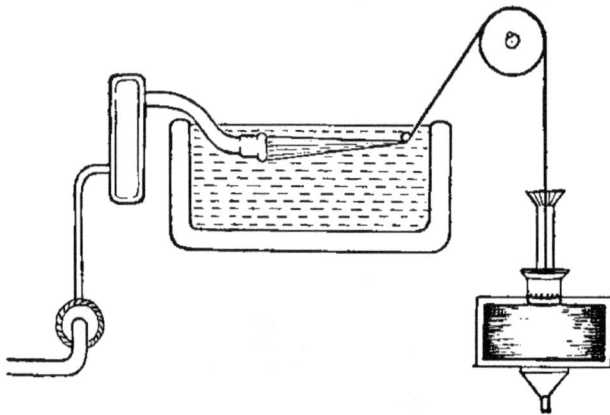

A spinneret, an acid bath, and a rayon thread

The crumbs are given various chemical treatments. At length a solution is obtained that is quite sticky and is ready for the next step in the process.

This sticky liquid is forced through the spinning machine (spinnerets) into a hardening bath. And then, at last, we have the solid glossy fiber coming out of the bath. It is ready to be sent to factories and used in making automobile robes, bedspreads, blankets, braids, carpets, coats, cravats, cushions, dresses, draperies, embroideries, and so on down the alphabet as far as underwear, for, woven alone or with cotton or silk or woolen yarns, rayon has come to be used to the extent of many millions of pounds yearly.

The Name *Rayon*

For many years the chemical man-made yarns and the articles made from them had no name of their own. All the other textile materials (materials that can be woven), like cotton, linen, silk, and wool, had names, but the new textile material was without a real name.

It was commonly spoken of as "artificial silk," a term that is still often used. It does resemble silk, having a silklike sheen. But it cannot truthfully be called silk, . and it seems unfair to call it artificial.

At last the Silk Association of America held a naming contest for the nameless material. People who were interested voted for a name for it. The name *rayon* was chosen because some one said the glistening stuff shone like light *rays*.

Weavers, knitters, dyers, and storekeepers welcomed this name. It was adopted by manufacturers in 1924. The Government of the United States recognized the name in an official way the next year. Dictionaries and encyclopedias published since that time have put the new word on their pages.

So, when you read about rayon in this book (and in many other places), you will understand that the word means the textile materials made from cellulose by any one of the four processes which were mentioned on page 327.

It may interest you to know, however, that the people who manufacture rayon have not all remained satisfied with this name. Some still use it. Others have chosen trade names of their own that they like better. If you notice advertisements of these textile fabrics, you will see some of the recent names given to certain kinds of rayon.

QUESTIONS AND ACTIVITIES

About how long is a fiber on a cotton seed? A fiber in a flax stem? A fiber unwound from a silk-worm's cocoon? Explain why a rayon fiber may be made any length the manufacturer wishes.

Ask permission to visit a pulp mill, if you live near one. Ask permission to visit a factory where rayon yarn is woven into garments, if you live near one.

Try to find out if any of your own clothes have rayon fibers in them.

Look for pictures of rayon articles in catalogs or magazines.

Plan a rayon exhibit at school, if your teacher is willing.

FUR AND WOOL

Clothes for Warmth

Eskimos dress themselves in the skins of furry animals. Air does not pass through these skins easily. They keep in the warmth of the body. They keep out the freezing cold. Clothes which prevent warm and cold air from passing through them are necessary for people dwelling in arctic regions. These people have to use the skins of animals for this.

Many people living in the north temperate zone need to work out of doors in winter weather or to travel where they are exposed to bitter winds. They can keep comfortable if they wear woolen underclothes and suits of hide or sometimes fur or fur-lined coats.

Courtesy Science Service, Washington

Eskimo hunter dressed in fur clothes

Aviators are exposed to cold when they fly high and fast. They have usually worn

sheepskin suits with sometimes a bit of
longer fur at the neck. Recently a new sort

Aviator dressed for winter flying

of flying suit has been designed. It is made
of calfskin with a collar of lamb fur.

Fur for Show

Most people who live in temperate climates and all people in tropical countries have no actual need of fur. Many of them, however, wear it.

Fur is used as trimming on collars of coats for men and women even though these collars may never be turned up at the neck for warmth. It is not uncommon to see people in thick fur coats riding in overheated street cars or busses when they would be much more comfortable in coats of a different sort. Evening gowns trimmed with fur are worn in ballrooms and at theater parties where the temperature is between 70° and 80°. A large and expensive fur scarf is sometimes seen during bright and sunny summer weather, when the wearer could not possibly suffer from cold without any scarf at all.

You will understand that the Eskimo and the aviator wear animal skins for needed warmth.   It is not hard to explain the custom of wearing such skins in cold climates in winter.

Do you know what reasons many people have for wearing fur which they do not need for warmth? Some wear it because they consider that it is becoming to them.   They think they look well in fur garments or fur trimmings.   Great numbers of people wear it because it is "fashionable" — so many other persons dress in fur that they follow the same custom.   In regard to "styles," people are often like a flock of sheep.   (Sheep follow a leader without giving thought of their own as to where they are going.)

Trapping Fur Bearers

Usually there is no reason to interfere with fashions or styles in dress. No one need worry whether women wear short hair or long hair.   It is not an important matter whether men wear knickerbockers or long trousers. There are two reasons, however, why many people have worried about much of the fur that is worn.

One reason is that, in order to supply the demand for costly and fashionable furs, many of the wild fur bearers have been killed. There is, indeed, grave danger that certain species of these interesting creatures may be exterminated from the earth.

This is a real grief to those people who believe that wild animals (except certain dangerous species) should be protected by mankind instead of destroyed in large numbers. They believe in *conserving* (saving) wild animals from wholesale destruction. They are called *conservationists*. Sometimes nations unite in efforts to save wild animals. In 1912 a fur-seal treaty was signed by England, Japan, Russia, and the United States. This treaty is still in force.

It is most important in conserving the seals
that make their homes on the Pribilof Islands
because it provides for their protection in
certain ways.

A second reason for objecting to the fur
trade, as it has been carried on for many

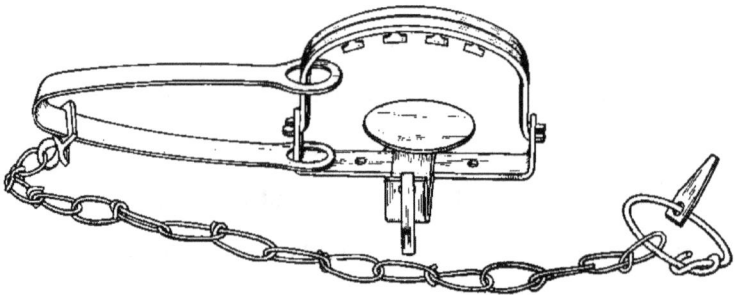

Steel traps like this cause suffering.

years, is that millions of the wild animals
taken by trappers each year are cruelly
caught in steel traps.

Animals caught in such traps suffer from
two causes. The trap itself breaks the vic-
tim's leg or hurts it otherwise. In addition
to this pain, the animal often starves before

the trapper comes to kill it. There are people who object to such cruelty. They say that animals killed for their fur should be killed quickly — that they should not be left in traps to suffer the agony of pain and starvation.

Humane Fur Laws

Cattle, sheep, pigs, and fowls whose flesh is used for food by people are killed quickly. There are laws that regulate the proceedings in slaughter yards. People who object to the use of painful traps in the fur trade believe that there should be laws preventing the sale of fur caught in this manner. They say that animals caught for fur should be killed as quickly as those used for food.

Laws against taking animals for fur in traps that torture have been passed in a few states. Many persons will not wear any fur that has been secured in a cruel manner.

Kindness seems more important to them than fashions.

Kindness to Animals

If you are a Boy Scout, you will know what the Sixth Law of your organization is. If you are not, you may be interested to ask some boy who is a member to tell you about this law.

We mention the Boy Scouts in this chapter because in certain places in this country they have played a part in a campaign to outlaw steel traps. Members of their troops of one district helped plan an exhibit for display at the Children's Fair at the American Museum of Natural History in New York. This exhibit showed why it is cruel to use steel traps.

You will be interested to know that the man who worked with the Boy Scouts on this exhibit was once a trapper himself. He gave up trapping when he realized how much pain he was causing to helpless creatures.

Grey Owl

You may have heard of Grey Owl, an Indian hunter and trapper who also found that he loved wild creatures too much to have further part in the wretched business of exterminating them. During recent years he has done all he could to protect the wild life of Canada.

An officer of the National Parks of Canada wrote us about how Grey Owl befriends wounded and deserted animals. This officer permits us to quote a part of his account of two young beavers that Grey Owl kept at his home one winter because they needed care:

"With the coming of spring both animals were returned to the water, where they soon took up their natural beaver life, repairing an old dam, felling trees and building their cosy home after the fashion of their kind.

Courtesy National Parks of Canada

Grey Owl befriends a baby beaver.

"Grey Owl has no desire to domesticate them or turn them into pampered pets; yet they continue fast friends. When his voice is heard calling at the landing place, they swim down the lake to him even if they are half a mile away.

"When he has been away on journeys out on the trail they often come to meet him, greeting him with little wiggles and squeals of delight. Or they may be waiting at the cabin door, eager to know what edible gift he has brought them.

"They like apples especially, and as he loosens his pack they will tug at the cords in an effort to help him open it. Then they will examine each package with almost childish curiosity, squealing with pleasure when they come upon their favorite fruit. Tearing open the bag, they will clutch as much as both paws can hold and stagger off to conceal their booty. Later they eat their treat, one apple at a time."

Fur Farms

Fortunately it is possible to obtain fur in humane ways. It is not necessary to take

it by trapping wild creatures. Foxes, rabbits, and other fur bearers reared on farms
may receive as good care as cows or sheep or

Photo by Lynwood M. Chace

A young fox

pigs.   When it is time to take their fur, they
are not left to suffer in traps but are killed
quickly.

As we just read, many kinds of fur bearers are already reared on farms (or ranches). It seems probable that as time goes on people will expect to have all their fur

From Ashbrook: "Fur Farming for Profit"

Rabbits with soft white fur

from ranch animals, just as they have their beef and mutton and pork from domestic animals.

The flesh of foxes is not sold for food. From rabbit farms, however, the flesh is sold in the meat market and the skins are sold in the fur market.

Karakul sheep have been tended by men for more than three thousand years. They were domestic animals even in ancient times.

From Ashbrook: "Fur Farming for Profit"

Karakul lambs

The fur of these lambs is sold under various names, such as Persian lamb, astrakhan, and Krimmer. Persian lamb and astrakhan furs are black, and Krimmer is gray. They are all soft and curly.

Old karakul sheep have fleece that grows to be from six to ten inches long. It is sheared and used for carpet wool.

Almost all the karakul fur that is worn in America is brought here from other countries. Some karakul sheep, however, have been brought to the United States and a few people are now rearing this kind of fur bearer here.

Wool

The fine, soft hair which forms the fleecy coat of certain animals is called wool. It is taken from the bodies of these animals without injuring them. So wool may be taken from the same animals each year. They are more comfortable without such thick coats during the summer.

Warm and durable woolen garments are made from the fleece of *camels*. A camel's longest hair grows on the neck and back.

This is usually pulled out in tufts at the right season of the year, when it is loose and ready to be shed. This longer overhair is

Photo by James C. Sawders, from R. I. Nesmith and Associates

Alpacas in Peru

rather coarse, but the camel has also soft, fine underhair. This fleece is cut like the wool of sheep. Cloth made from it is soft and light.

The *llama* and *alpaca* are relatives of the camels. They live in South America. The wool from the alpaca makes excellent cloth.

© *Ellison Photo Company, Austin, Texas*

An Angora goat in Texas

Llama wool is coarser than alpaca wool, but some people use it for clothing and blankets.

The wool of goats is used for many purposes. *Angora goats* have long, strong, shiny hair. One name for it is *mohair*.

The fleece of Angora kids (young goats) is fine and curly. Angora wool is used for the glossy cloth called *brilliantine* and for sweaters and outing suits. Fuzzy hoods for children are made with it. The soft plushes, covering the cushions in railroad cars and automobiles, are mohair products.

At one time Angora goats could be found only among the hills of Asia Minor. But goats of this sort have been introduced into other countries, and now they may be met in many parts of the world. More goats are reared on Texas ranches than in other places in the United States.

Most of the woolen cloth in the world is woven from the fleece of sheep. This is sheared from the bodies of the living sheep each year. It is also taken from sheep that are killed for meat.

Soon after white people began to settle in America, sheep were brought here from other countries. More and more of these animals were reared here, until at one time there were more than fifty million sheep

Courtesy U. S. Dept. of Agriculture

Ewe (mother sheep) and lamb

feeding on grassy fields in the United States. Most of them were on western plains. No other country had so many sheep. Now, however, there are not so many sheep in the

United States as there are in Australia or
Russia.

If you look at one of your own hairs
through a microscope, you will find that it

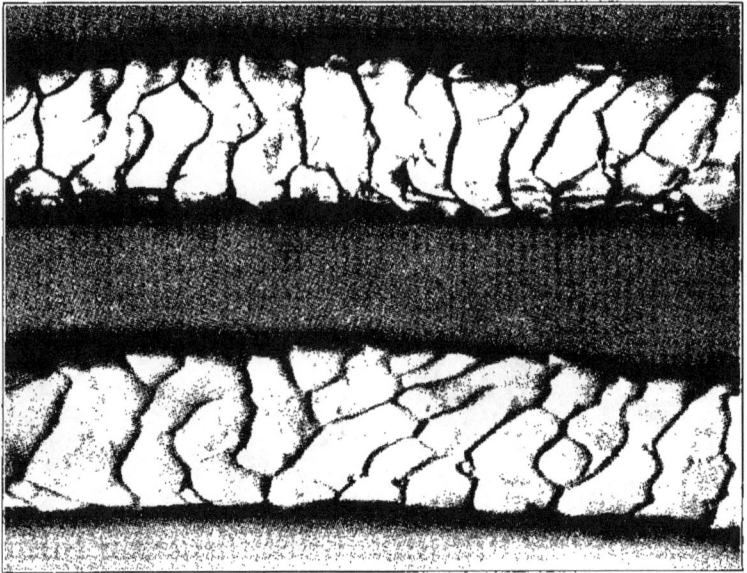

Photo by J. Manley, courtesy U. S. Dept. of Agriculture

Two wool fibers enlarged

Notice the overlapping scales on the fibers.

has a smooth surface. Most kinds of mammals have smooth hairs. Wool fibers,
however, are covered with hundreds of tiny

overlapping scales. These scales are so small that you cannot see them unless you look at them through a microscope. When the wool is spun into yarn, the scales on one hair catch and tangle into those of other hairs. In this way the scales hold the fibers together and make the yarn strong.

Questions and Activities

Take a few of your own hairs. Twist them together. Do they tangle in such a way that they make a strong yarn or thread?

Can you get some fleece from a sheep? If so, twist a few hairs together. Do they cling to one another better than your hair did?

Ernest Thompson Seton said, "Trapping wild animals with steel traps is a wretchedly cruel business and will doubtless be forbidden by law before long." His book in which this remark is made is called *The Book of Woodcraft*.* If there is a copy of this book in the library you visit, look in the index for "Trapping animals" and find the page

*See page 424 for book list.

Sheep grazing on land in a national forest

on which he wrote about catching animals without doing them any injury. Notice the picture of the "ketchalive" trap.

Did you ever hear of the Anti-Steel-Trap League? This is a national league formed to outlaw the steel trap as a means of getting fur. Dallas Lore Sharp was one of the vice presidents of this league. Four of his nature books are called *The Spring of the Year*, *Summer*, *The Fall of the Year*, and *Winter*. Try to get a copy of at least one of them to read or examine.

What reasons could Dallas Lore Sharp and Ernest Thompson Seton have for objecting to steel traps?

What is the Sixth Law of the Boy Scouts?

If *Holiday Pond* is in your library, read the chapter called "Lotor, the Washer." After you have read this, tell what was on the sign the farmer nailed to a tree.

What would happen to the fur business if fur should become unfashionable?

On page 360 there is a picture of sheep on land of the Mt. Hood National Forest, Oregon. Are these karakul sheep? How can you tell whether they are karakul or some other kind?

Write a short essay using either "Fur from Sheep" or "Wool from Sheep" for your title.

Look at a map of the Bering Sea, west of Alaska, and find the Pribilof Islands.

FOOD

PART
SIX

Eating some "fuel" for luncheon

FUEL FOODS

While reading this book you have been thinking about homes as you would about museums — you have been a bit curious concerning the exhibits you find.

First you read about gardens. Next you learned about different materials of which the shelters themselves are made. Then you turned your attention to such important subjects as water, light, and heat. Later you noticed common household tools and appliances and learned how they are used. After that you visited clothes closets and read about materials of which most of our clothes are made.

Now it is time to think about the substances you eat. In the next few chapters you will be reading about food. You may learn new names for the food you eat every day. It is quite likely, indeed, that you will begin to think of foodstuffs in ways that are new to you. For example, you may not be in the habit of calling some of your food *fuel* as if, in some way, it were like coal that is fed to a railway engine or gasoline that is poured into the tank of an automobile.

Fuel for Heat and Other Energy

What becomes of fuel that is used to make a machine do its work? We say such fuel *burns*. As you have read, there is a substance called carbon in coal and wood; it is in gasoline and other fuels we use in machines. You also know that there is oxygen in the air. When the fuel burns, carbon and oxygen unite. While the carbon and oxygen are uniting, they furnish *heat* and may furnish other forms of *energy*.

We do not always value both the heat and the other energy. When we burn wood in a fireplace or coal in a cookstove, we do so for the sake of the heat. When we burn coal in a railway engine or gasoline in an automobile, we use energy to produce motion.

You may think of your body as a sort of living machine. It needs fuel to make it warm as much as a furnace needs it to make a house warm. It needs fuel to give it energy to move as much as an automobile or a steamboat does.

The fuel your body needs must have the same important substance in it that is burned in other machines. You must have carbon. Of course you do not shovel coal into your mouth or pour in gasoline. But you do eat other kinds of fuels that have carbon in them.

You take carbon into your body when you eat fuel foods. You take in oxygen when you breathe. Carbon and oxygen unite inside your body. While they are uniting, they supply heat and other energy. The heat keeps your blood warm. Some of the energy gives you power to move — to run and play and work. It also keeps your heart beating and other parts of your body going inside.

Without fuel a steel engine becomes cold and still. Your body would also lose its warmth and power to move if you did not give it fuel.

There are three classes of fuel foods. Sugars and starches belong to one of these classes. Fats form a second class. Lean meat and some other foods are in a third class.

Sugars and Starches

The green material that is in the leaves of leafy plants is called *chlorophyll*. There are enormous numbers of tiny green particles in growing leaves. It is chlorophyll, too, that makes the stem of a cactus plant green.

All plants that have chlorophyll can make sugar. They are sugar factories that run by sunlight. The materials they use are water and the gas called carbon dioxide.

As you know, there is always water in living leaves, since it runs through the plant in the form of sap after the roots have taken it from the soil.

The supply of carbon dioxide is present in the air. Whenever a fire burns, some carbon dioxide escapes in the air. Whenever an animal breathes or a plant respires, some of the same gas is given off. Whenever substances decay, gas of this sort comes away from the rotting material. There is not much of this gas in the air, but there is enough for plants to use in making sugar.

As the name, *carbon dioxide*, suggests, there are two substances in this gas. One is carbon. The other is oxygen. Carbon dioxide is formed when one part of carbon combines with two parts of oxygen. If you wish to write the name of this gas in a short way, you may use the initial letters of the two substances that are in it. You may call it CO_2. C stands for carbon. O stands for oxygen. The figure 2 near the O means that there are two parts of oxygen.

There are also two substances in water. One is hydrogen and the other is oxygen. The recipe for water is two parts of hydrogen and one part of oxygen. So H_2O means water.

Plants take H_2O out of the ground and CO_2 out of the air. Then they combine carbon and hydrogen and oxygen in just the right way to make sugar. The symbol for ordinary sugar is $C_{12}H_{22}O_{11}$ since it is a combination of twelve parts of carbon, twenty-two parts of hydrogen, and eleven parts of oxygen. (Men have recently learned how to make sugar from these three elements, but this artificial sugar is too costly for common use.)

Some plants have no chlorophyll. Mushrooms and other fungi are such plants.

Plants without chlorophyll cannot make sugar. They resemble animals in that they could not live if all green plants were to perish. The sugar of the world is made by plants that have green chlorophyll. They run all the real sugar factories. But, as we said, these factories work only in the sunlight. They cannot work "overtime" or in "night shifts," although they *do* work steadily during the short summers of arctic regions and they *can* work by some kinds of artificial light.

Of course you have heard of places called sugar factories, where men prepare sugar (or sugary sirup) from sugar cane or sorghum or beets or sap from maple trees. But the men in such factories do not make the sugar itself. They merely prepare it for use.

The same is true of the sugar present in the bodies of certain insects. The insects do not make the sugar. They take it from plants. Honey, for example, is a form of

Bees after nectar and ants after honeydew

sugar sirup. Bees do not make the sugar in honey. As you know, they gather it in the nectar they drink from flowers. Honeydew is another form of sugar sirup. It is a sweet liquid that comes from the bodies of aphids and some other insects that are related to aphids. These insects do not make the sugar in honeydew. They find it in the sap they suck from plants. Ants like to drink honeydew. You may have found them visiting aphids for honeydew as bees visit flowers for nectar. Bees, too, sometimes sip honeydew, but they make better honey from nectar.

There is a different kind of sugar in the milk of cows and other mammals. It is called *milk sugar*, and mammals make it from other fuel foods.

Plants, as you know, use sugar that they make for food. As sugar dissolves in water, it is not easy for plants to keep all they need

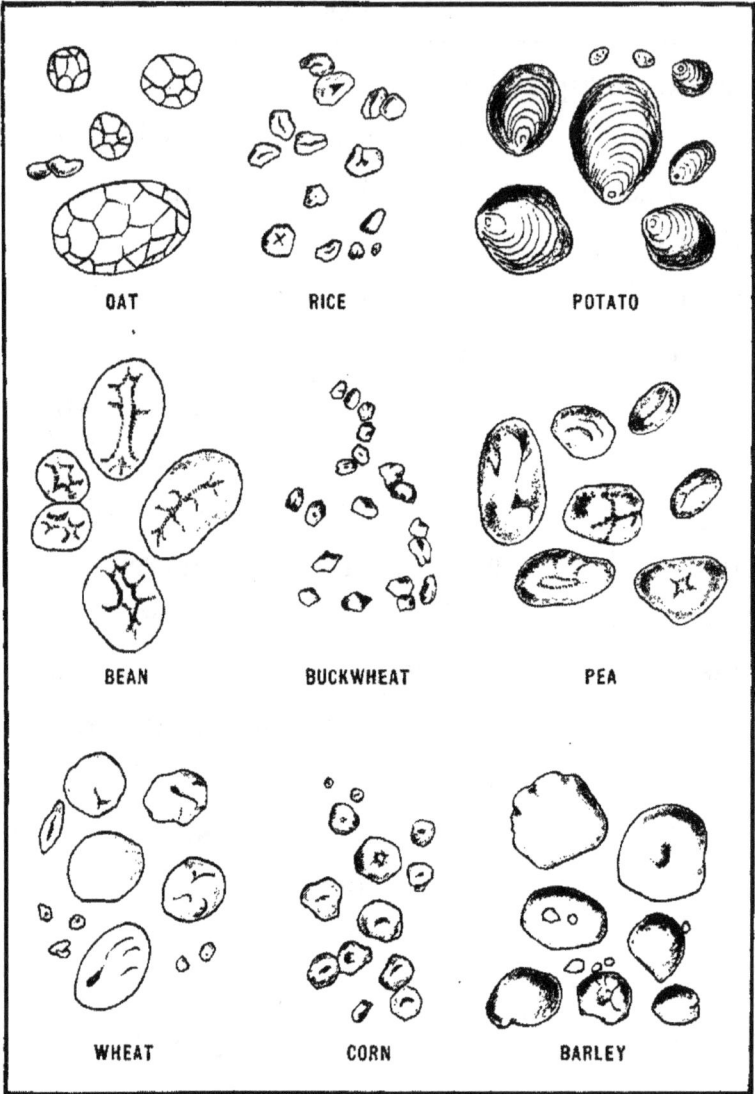

How starch granules of some plants look when seen
through a magnifying glass

in this liquid form. They change most of the sugar they make into starch, which does not dissolve in water. It remains in tiny solid particles. Plants store the starch until it is needed for food. Then they change it back to sugar as easily as they changed the sugar to starch. Starch and sugar are both made of the same three substances — carbon, hydrogen, and oxygen.

Each plant has some way of storing starch particles, or *granules*. People who have studied starch granules by looking at them through microscopes have learned that the starch granules found in plants of one species differ in shape from those found in plants of other kinds. Thus a granule that a rice plant stores in its seed has not the same shape as a starch granule stored in a potato tuber or as one stored in an onion bulb. Even in the same plant family, each different species has its own sort of starch granule. For example, the starch granules in a bean seed are not like those in peas or peanuts, although beans, peas, and peanuts are all plants belonging to the Pulse Family.

Sugar, as you know, tastes sweet to you as soon as you take it into your mouth. Starch, however, does not have a sweet taste at first. But if you will chew a piece of wheat bread slowly, you will notice that this sort of food becomes sweet in a short time. It begins to taste sweet as soon as the saliva in your mouth has begun to mix with it. Saliva changes some of the starch in your food to sugar. After you swallow your food, even more of the starch is changed to sugar while it is being digested.

Sugar (with starch which is changed to sugar) is a most important food for you to have. The carbon in it makes it an excellent fuel food. It gives you warmth and other kinds of energy.

You will do well to realize, however, that it is possible to have too much of even a good thing. You will find that it is an excellent plan to take the advice of older people about the amount of fuel you eat in the form of candy and other sweets.

In another chapter ("Rayon") you learned that leaves, woody stems, and other parts of plants contain certain fibers of a substance called cellulose. The celluloses of wood are examples. Cellulose is related to the starches and is derived from the sugars.

Some herbivorous (plant-eating) animals, it is true, can thrive upon various sorts of shoots and leaves even when they are "cured" as hay. It is also true that certain molds can attack the purer cellulose of moist paper or of moist cotton cloth. "Greens," however, nourish us human beings almost entirely by their juices or saps.

The bulkiness of some plant parts that we are unable to digest is a benefit to us. Cooking often helps both to change the condition of these bulky parts and to release and improve the flavor of the juices. Cooking serves also to break up and soften the starch granules.

The "colored water" left in any vegetable kettle often contains the most nourishing part of the plant material. In boiling or steaming vegetables therefore it is important to avoid using too much water, thus thinning the juicy liquid more than is necessary. People who know the food value of this liquid do not waste it by throwing it away. They use it in soups or combine it with foods in some other way.

Fats

You may have heard some one say, "The fat is in the fire." You may know that fats

burn easily.   If they are thrown into a fire,
they blaze quickly.

There is carbon in fat.   When the fat
burns, some of the carbon combines with

The fat is in the fire.

oxygen to form carbon dioxide (CO_2).   Fats,
as well as sugars and starches, are fuel foods.

After you eat fat, it burns in your body.
It does not make a flame but it does produce
heat.   It also supplies other kinds of energy.

It helps your muscles and organs do their work. It helps keep you strong and able to move easily.

We take some of our fat foods from animals. Most kinds of meat taste better if there is at least a little fat cooked with them. "A streak of fat and a streak of lean" is the way most people like their ham and some other meats. The fattest part of a hog is separated from the rest of the body and prepared as lard. Suet and tallow are fatty portions taken from the bodies of cattle and sheep.

There is only a little fat in skimmed (or separated) milk. Cream is very rich in fat. When cream is churned, the fat collects in solid lumps. These are used as butter. The liquid part of the cream that is left in the churn is called buttermilk. Another food that has fat in it is the yolk of eggs.

Besides many different animal fats which we eat, there are also fat foods that we get from plants. Olive oil, cottonseed oil, and other vegetable oils which we use with salads are fats that we often eat. There are fats in nuts, too.

Proteins Are Fuel Foods

You have read about two classes of fuel foods — sugars (and starches) and fats. Now you come to the third class, *proteins*.

You take protein food whenever you eat meats (including fowl, fish, and shellfish). Eggs and milk also contain various protein foods, as in the *albumen* of egg whites and the *casein* of milk curds.

There are proteins in plants as well as in animals. Nuts are rich in protein. So are the seeds of beans, peas, and other members of the Pulse, or Pea, Family.

You have already learned that milk has a special kind of sugar, called milk sugar, and that it has fat. Now you are learning that it also has protein. Many other foods have all three of these fuels in them. The next time you eat some peanut butter you might say, "I am eating starch, fat, and protein (three kinds of fuel foods) all at once!"

Since proteins serve, among other uses, as fuel foods, you will guess that they, too, must have carbon in them. So they have. When these proteins burn in your body, some of the carbon unites with oxygen to form carbon dioxide (CO_2). This of course is what happens also when sugars and fats burn in your body. It would happen, too, if you burned any of these foods in a stove or on a bonfire.

Proteins, however, have several substances that neither sugars (and starches) nor fats have. When you read the next chapter, you will learn what one (the most important) of these different substances is.

Questions and Activities

People who have hard labor to do need more energy than people who are sitting quietly almost all day. Which people should eat more fuel foods?

Chew an unsweetened cracker slowly. How does it taste when you begin to chew it? How does it taste just before you swallow it? Why is there a change in the taste of the cracker? In this chapter (page 376) you read that even more of the starch is changed to sugar after you have swallowed starchy food. What happens when the hydrochloric acid in your gastric juice comes in contact with starch in your stomach? If you cannot answer that question, read pages 32–34 of *Through Four Seasons.**

If one of your friends said to you, "Please give me a drink of H_2O," what would you give him?

Peel a raw potato. Cut the peeled potato into thin slices. Let the slices remain for a few hours in a glass dish with water enough to cover them. Then remove the slices from the water. You will notice that the water in the glass looks rather milky. This appearance is caused by the large

* See page 424 for book list.

number of very small starch granules that have come from the potato. Let the glass of whitish water stand for ten or twelve hours. Is the water still milky looking or is it clear? Did the starch granules rise to the top of the glass or sink to the bottom?

What happens to sugar when you leave it in water?

You have read that starch granules do not dissolve in water. They do not even change their shape in cold water. Heat, however, makes them swell and burst. What happens to the starch in a kernel of pop corn when it is heated? Why is a popped kernel larger than an unpopped one?

Ask if you may make this experiment: Stir a tablespoonful of cornstarch in a cup of cold water. What happens? Next, boil the water and cornstarch. What happens then?

BODY BUILDERS

Your body must be built just as surely as a house is built. You cannot see what is taking place in your body from day to day, but you do know that you become taller and heavier. Perhaps friends of your family say to you, every now and then, "Why, how fast you grow!"

Many foods that you eat have things in them that help a little to keep your body growing. Some foods, however, have materials that do more than any others to build your body bigger and stronger. Such foods are sometimes called "body builders."

All things become worn in some way or another. You know how your clothes must be mended. Then, too, you know that automobiles must be taken to a garage for repairs.   Worn parts in our houses must be mended or replaced.

You may not have thought of it, but your body needs to be mended (repaired) every day.   Indeed, the repair processes are taking place in your body constantly — every hour, every minute, every second.

The same materials that build bodies are those that keep them in repair.   Cells are always multiplying and pushing out in different directions to make larger bodies for growing boys and girls.   And all the time worn-out cells are being replaced by new cells.

In this chapter we shall speak of three classes of body builders — proteins, minerals, and water.

Proteins Are Body Builders

In the chapter before this we told you that proteins are fuel foods.   In this chapter we shall speak of them also as body builders.

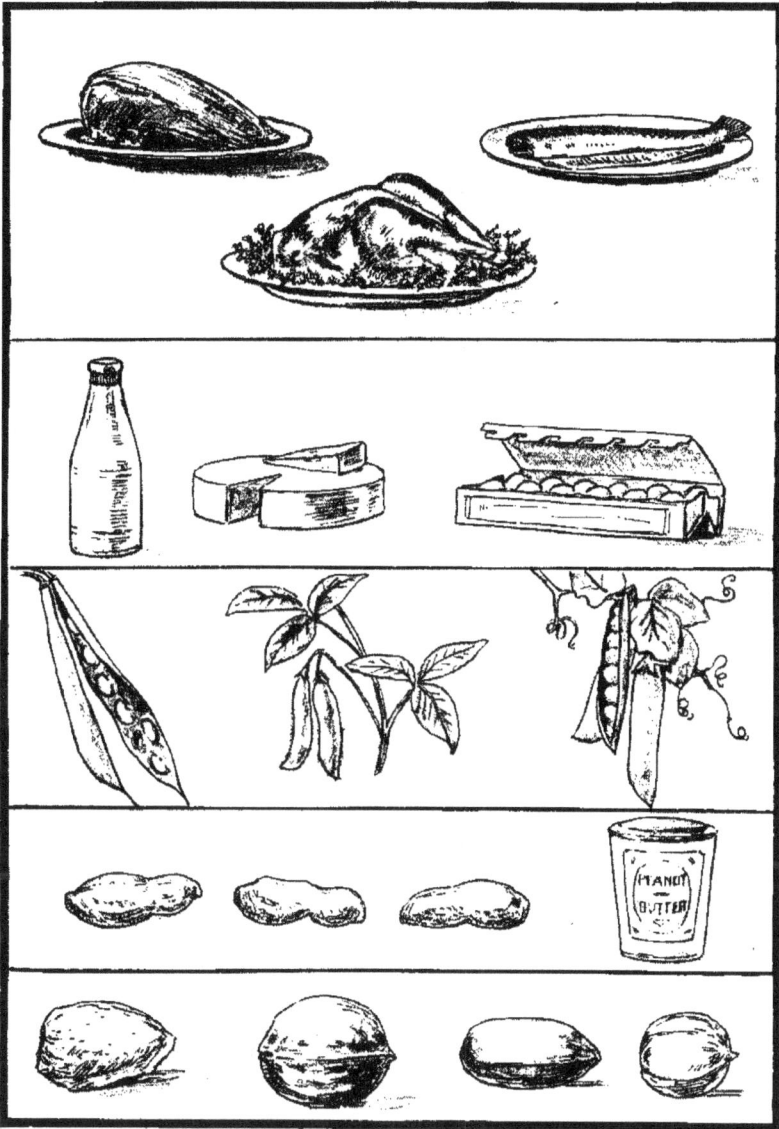

Some foods with proteins in them

Sugars (and starches) and fats have these three different chemical substances in them — carbon, hydrogen, and oxygen. Proteins have the same three substances. They also have nitrogen, which is a substance not present in sugars and fats.

Perhaps you already know something about nitrogen. If you read *Through Four Seasons,** you may remember studying about nitrogen and learning how hungry plants get it for food.

Every tiny living cell in the body of every plant and animal must have some nitrogen. Nothing can be alive without it.

Nearly four fifths of the air we breathe is nitrogen. We are surrounded by nitrogen all the time, though we can neither see it nor smell it. You might think that your body could get all the nitrogen it needs by breathing. But, strangely enough, you cannot

* See page 424 for book list.

use any of the nitrogen you take in this way. You breathe out all the nitrogen that you breathe in.

(About one fifth of the air is oxygen. If the air had as much oxygen as it has nitrogen, all the flaming fires in stoves and lamps and other places would burn with great violence.)

Since we must have nitrogen and cannot use what we take into our lungs, we must depend on food for our nitrogen supply. The nitrogen in protein foods is the same substance as the nitrogen in the air. But in these foods it is combined with other things in such ways that our bodies can get hold of it and use it.

Many foods have some proteins in them, but those that give us most of the nitrogen we can use were mentioned in the chapter on fuel foods — lean meats of all kinds, eggs, milk (and cheese and other foods made from milk), and seeds of some plants.

Among the seeds that are rich in usable nitrogen are nuts and legumes. *Legume* is another name for a plant belonging to the Pulse, or Pea, Family. Peas, beans, and lentils are seeds of certain of these plants most commonly used for food. Peanuts, as you know, are also seeds of a plant belonging to this same family.

On another page, we remarked that it is quite easy to take too much of even such a necessary food as sugar. We may say here that we do not advise you to eat a great many nuts just because they are rich in proteins. The proteins in nuts are so hard and firm that they cannot be digested unless they are ground into fine bits. One way to grind them is with our teeth. Nuts should be chewed very thoroughly indeed. Some nut foods, such as peanut butter, are ground by machines so that they are finer than they can be chewed. That is the reason that

children can often eat a peanut-butter sand-
wich much more safely than they can eat
nuts that have not been ground by machinery.

Some Minerals Are Body Builders

What happens when you burn a stick of
wood? Most of the wood disappears. All

Nothing will be left but ashes.

that remains is a little pile of ashes. These
ashes are the mineral parts of the wood that
are left when all the rest of the wood is gone.

There are some minerals in all plants and animals. There would be ashes left if any one of them was burned.

If the body of a man weighing one hundred fifty pounds were to be thoroughly burned, there would be about six pounds of ashes left. These ashes would contain nine different minerals that had been important in building and repairing that man's body. They are commonly called *mineral salts*.

You have the same nine minerals in your own body. Four of these will be mentioned in this chapter. In eating foods that have these four minerals, you get the other five also.

Two Minerals in Bones

There are two minerals that have most to do with the hardness and strength of our bones. One of these is *calcium*. You know something about lime and limestone, but perhaps you have not learned that they have calcium in them.

It is this calcium that makes lime a good food for plants. (Lime also makes soil less acid.) Farmers often spread lime on their meadows. It is washed into the soil by rain. The grass plants take it from the soil through their roots.

A cow that eats the grass gets the calcium into her body. Some of it is used in keeping her bones and teeth in good condition. Some of it is put into her milk.

When you drink milk, you get calcium in a form that is easy for your body to use in building and repairing your bones. You also eat calcium with your vegetable food — as the cow does. There is more calcium in your body than any other mineral.

Calcium alone, however, would not strengthen your bones and harden your teeth. The other mineral that helps is *phosphorus*.

You may remember reading about phosphorus in the book *Through Four Seasons.*
It is another substance that plants need for

Courtesy U. S. Dept. of Agriculture

Cows get calcium when they eat grass.

food. There is phosphorus in chemical fertilizers that may be bought to put on the soil for plants.

* See page 424 for book list.

There are many uses for this chemical. It is one of the substances in the mixture which forms the head of a match. It is the phosphorus which causes the first flare when you make the match head hot by rubbing it on a rough surface.

You get calcium and phosphorus when you eat cabbage, celery, squash, spinach, lettuce, asparagus, and other vegetables.

Iron

A full-grown man has enough iron in his body to make a good-sized nail. That is not much, but it is very important. Iron is needed especially in the blood.

There is iron in many kinds of food. Milk has only a little iron, but it is easy for the body to use it. There is more iron in lean meat, oatmeal, wheat (whole grain), spinach, beans, and peas. Egg yolk has still more iron than these other foods.

Iodine

Iodine is another necessary chemical mineral for a healthy body. We need even less iodine than iron, but that little is important.

Pure iodine is a dark purplish hard crystal. In this pure state it is a poison. You may have seen a dark liquid called tincture of iodine. This is a liquid in which the iodine crystals are dissolved. In this form iodine is also a poison and must not be swallowed. It is, however, good to use on cuts and some other wounds to clean them so that they will heal quickly.

Iodine is present in sea water and in animals and plants that live in the sea. There is a little iodine in some mineral springs. Vegetables that are grown in some soil, usually near the sea, have some iodine in them.

People living near the sea get all the iodine they need. Those living in some other places sometimes need to take small amounts of some compound of iodine as medicine. In some places it is added to table salt. In some cities it is put into the water supply.

People and other animals that do not get enough iodine are liable to have swollen glands in their necks. We say then that such a person has a *goiter*.

Water

When you empty the water from a plump rubber hot-water bottle, you have only a limp cover left. If a full bottle holding four pints of water weighs four and a quarter pounds, it would weigh one quarter of a pound when empty.

If you should leave a raspberry on a blotter in the hot sunshine until it is dry, there would be little left but the seeds and skin. The watery juice, which gave it most of its size and weight, would be gone.

It is easy enough to understand that tomatoes, watermelons, and oranges are mostly water. But perhaps you have never realized that potatoes and beets and other vegetables that seem solid have much more water than anything else in them, too.

Have you ever been told that you yourself are almost all water? If the body of a child weighing one hundred pounds were to become thoroughly dried, there would remain only about thirty pounds. It is, indeed, water that builds almost all the bulk of your body, just as surely as it is water that builds most of a potato.

You lose considerable water from your body every day. Some of this water escapes from the lungs in moist breath; some goes out through the pores of the skin in the form of sweat (perspiration); and some is removed from the body by the action of organs called the *kidneys*.

Of course you must keep up the supply of water in order that the parts of your body may be properly built and repaired. If you go without water, you become thirsty — your thirst is a signal for you to give your body water.

You get some water with most of the food you eat. Even bread, baked in a hot oven, has a little water in it. There is water in all vegetables and fruit. Soups, milk, and other liquid foods are mostly water. Besides all the water you take in such ways, however, you still need to drink water, by itself, in order to give your body enough to keep it in good condition.

QUESTIONS AND ACTIVITIES

In this chapter you have read about many substances, including lime, nitrogen, and phosphorus. Look for these three words in the index of *Through Four Seasons*.* Then turn to the right pages and

*See page 424 for book list.

read what is written about these subjects in that book.

Take enough of any raw fruit or vegetable (such as apples, oranges, tomatoes, beets, spinach, or potatoes) to weigh one pound.

Put the pound of vegetable or fruit through a meat grinder. Strain (or squeeze) out as much liquid as you can. Then dry the ground, pulpy mass in a hot, sunny place or in an oven (with the oven door open to let out the steam).

After you have dried the mass as well as you can, weigh it. Does it still weigh a pound? How much of your vegetable or fruit was water that has been removed by straining (or squeezing) and drying?

VITAMINS

In earlier pages of this book we spoke of the two chemical substances in water - - hydrogen and oxygen. We told you that sugars (and starches) and fats have these same two chemicals and one more — carbon. When we were talking about protein foods, we said that they had all three of the chemicals just mentioned and another that is very important — nitrogen.

This chapter is about *vitamins*, but we shall not tell you what chemicals there are in them. The reason we do not tell you is that we do not know. At the time this chapter is being written there are a number of things about vitamins that chemists have not yet learned.

Do you think it is foolish to have a chapter about substances when we cannot tell you

Some foods that have vitamin A

what is in them? Perhaps, after you have
read this chapter, you will be glad that we
did not leave it out of the book. You may
be interested to know what vitamins you
eat with certain foods even though you do
not know just what the vitamin itself is.
You may like, too, to learn which vitamins
help keep you well in some important
ways.

You will recall that you knew a great deal
about water long before you ever heard of
either hydrogen or oxygen. Well, chemists
have learned much about vitamins before
knowing what chemicals are in them.

Chemists often need new names for substances when they begin to study them.
They had a difficult time deciding what to
call the different vitamins. They gave them
long names at first. One of these names has
seventeen letters in it! But even the chemists themselves did not find these names

easy to use. After a time they agreed to speak of vitamins by letters instead.

So you will not need to be bothered with hard names in this chapter. You will be reading about vitamins A, B, C, and D. There are other vitamins besides these, but four are enough to study in this book.

Vitamin A

Vitamin A is found in many plants. There is more of it in green leaves than in other parts of leafy plants. You get some of this vitamin when you eat spinach or lettuce or other leaves. After an animal of any kind eats green leaves, some of the vitamin A goes to different parts of its body.

Hens need some green plant food. They store so much of the vitamin in the yolks of their eggs that we say egg yolks are rich in vitamin A.

Cows eat grass and clover and other green leaves. Some of the vitamin A they take finds its way to the milk glands and is stored in the milk. So you get some of it yourself when you drink milk. There is some, too, in cheese and butter and other foods made from milk.

The liver is an organ which holds more of this substance than most other parts of the body. If, for example, you eat beef liver for dinner, you get some of the vitamin A which that animal took when it ate green grass. So you can get some of this vitamin without eating any of the grass yourself.

In the ocean there are millions of tiny plants in which vitamin A is formed. They are eaten by very small fish and other little sea animals. All the creatures that eat the sea plants take vitamin A into their bodies.

Larger fish eat the smallest ones, and then they get vitamin A in this way. Fish store

"All the creatures that eat sea plants take vitamin
A into their bodies."

much of this vitamin in fatty (oily) parts of their bodies. Some fish have good supplies of oil in their livers. This is true of codfish and halibut. Vitamin A is stored in this oil.

So, when children take cod-liver oil or halibut-liver oil, they get vitamin A that was formed by sea plants. They do not need to eat the sea plants themselves. The fish save them the trouble!

For many years it has been the custom to take cod-liver oil that has been placed in bottles. It may still be taken in that way. Recently, however, scientists have discovered ways to prepare the vitamins that are in these oils so that they may be taken in the form of capsules or tablets.

Vitamin A is good for children because it helps their growth. Older people need it to aid them in keeping their bodies in good condition.

This important vitamin, too, prevents people from having one kind of eye disease. The Japanese learned long ago what to do to help children who had this disease. They fed these children livers of chickens to cure them. They knew there was something in liver that would prevent this eye disease, but they did not know what it was. Now, years and years later, we call this helpful substance vitamin A.

Vitamin B

For a long time men in the Japanese navy were troubled with a disease called *beriberi*. All those years their chief food was polished rice. At last some one tried to find out whether the health of these men would be improved by other food. They were given whole grain instead of polished grain to eat. Soon after that change in their food, the disease beriberi became very rare instead of a common trouble.

People who have studied this subject tell us that there is considerable vitamin B in the germs of cereals. (*Cereal* is a name given to a grain or a plant belonging to the Grass Family that has seeds good for food. The *germ* of a seed is the part that sprouts when the young plant begins to grow.) They tell us, too, that vitamin B is found especially in the *bran*, or outer portion of the grain. When rice is polished, the bran and most of the germ are removed. So people who do not eat much except polished rice do not have enough vitamin B in their food.

You may have heard that whole wheat made into bread or other cereal food is good for you to eat. One reason why this is so is that you get the vitamin B that is in the bran. The bran is removed from wheat kernels that are made into white flour.

This vitamin is found in various vegetables and fruits. Tomatoes, peas, asparagus, and beans are four foods that are regarded as supplying this substance in helpful amounts.

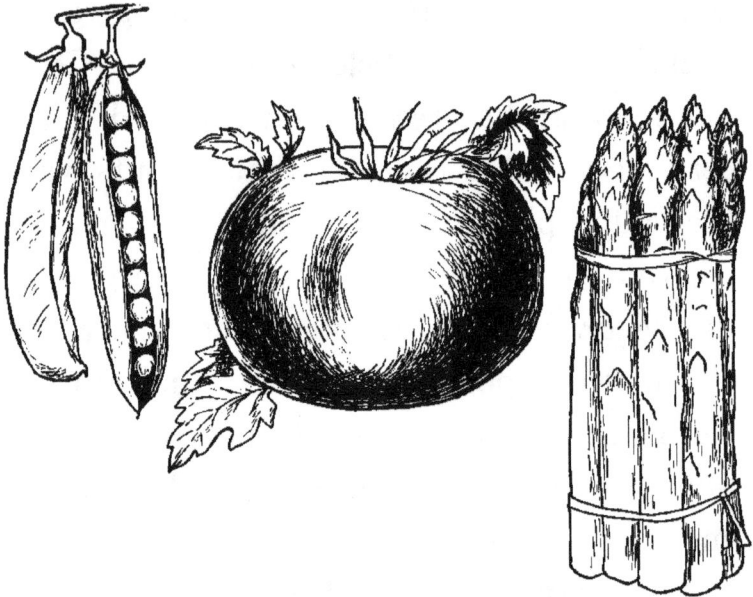

Some of the foods that have vitamin B

Vitamin B helps you to have a healthy appetite. It aids you to digest your food. It is necessary for proper growth.

Vitamin C

You may have heard of vitamin C. Perhaps you have read that oranges and lemons

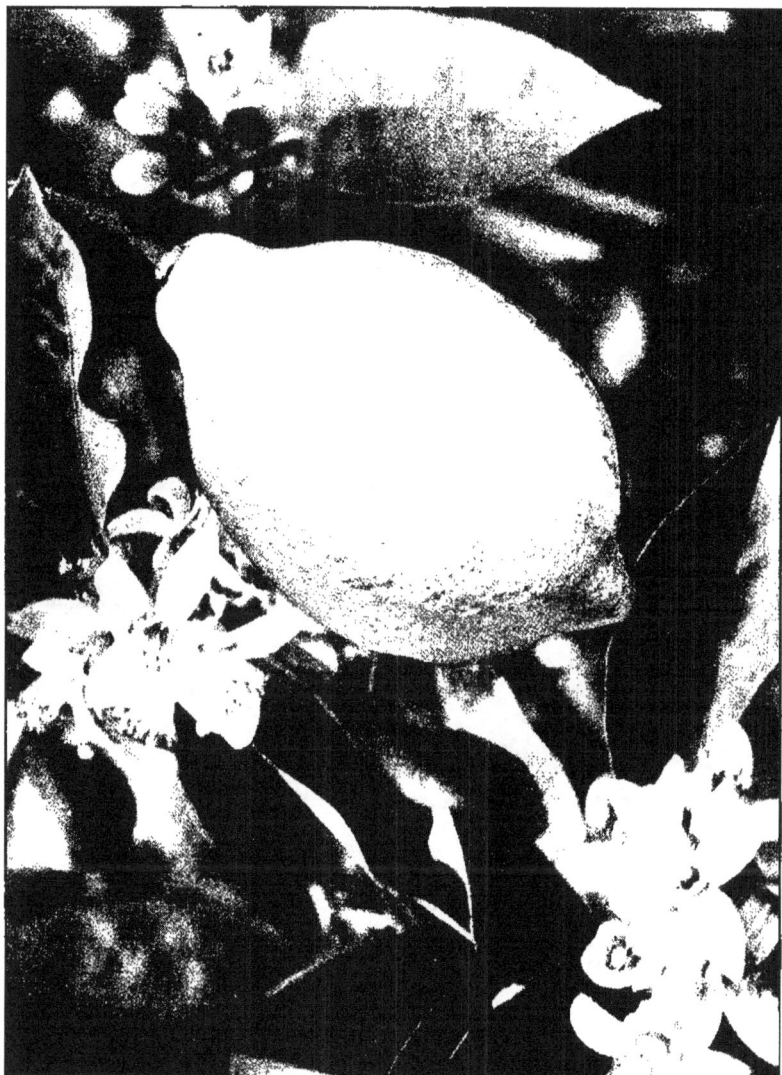

Fruit and blossoms of a lemon tree

and grapefruits have this vitamin. These three kinds of fruits grow on trees that belong to the Rue Family. They are called citrus

Courtesy California Fruit Growers Exchange

Oil-burning heaters in a citrus orchard

fruits. We shall speak of them first, and later we shall tell you of some other foods that have a supply of vitamin C.

Oranges and other citrus fruits grow only in climates where there are no cold winters. So we do not find any groves of these trees in the North. Even in California and Florida

Courtesy California Fruit Growers Exchange

Fumigation tents in a citrus orchard

heaters are sometimes used in the orchards to keep the fruit and trees from frost injury during winter nights.

If you read the book *Through Four Seasons,** you may remember about some of

* See page 424 for book list.

the insects that trouble apple trees. One of these insects is the oyster-shell scale. Another kind of scale injures citrus trees. So it is a common practice to cover the trees with canvas tents and fumigate them with a poisonous gas. If this is done at night, when there is no hot sun, the treatment does not injure the trees and it kills the scales and other insect pests. The trees are fumigated about once a year.

At other times citrus orchards are dusted with sulfur to kill thrips and certain other injurious insects.

All this care and much other work is necessary in order that people may have citrus fruit to enjoy.

Such fruit was known to be healthful long before anyone ever heard of a vitamin. More than two hundred years ago, indeed, a man who was writing about a disease called scurvy said that medicine could not cure anyone who was sick with it. "But," he wrote, "if you can get green vegetables, . . . or if you have oranges, lemons, citrons, or their pulp

Courtesy Paul S. Smith

Dusting a citrus orchard with sulfur

and juice preserved with whey . . . you will, without other assistance, cure this dreadful evil."

People who have plenty of fresh (or properly preserved) vegetables and fruit are not troubled with scurvy. Before this was known, sailors were not given the sort of food they needed. The old sailing vessels were out for many months at a time without carrying food that could save those on board from scurvy. Thousands of British sailors were sick with this disease every year. In 1804, however, a rule was made that every sailor should be given a daily allowance of lime juice. (Limes, as you may know, are the fruits of trees closely related to lemon trees.) After that these men found that they had no cause to fear scurvy.

Citrus fruits or their juices can often be carried more conveniently than other foods that have vitamin C. It is easier to keep them fresh than many other foods. These and other reasons cause us to value these important fruits very highly indeed.

Fortunately, however, it is possible to get plenty of vitamin C in other foods, too. People who have studied this subject tell us that apples, ripe bananas, cabbages, and onions are excellent foods in which to get vitamin C if they are eaten raw. Tomatoes and strawberries are rich in vitamin C

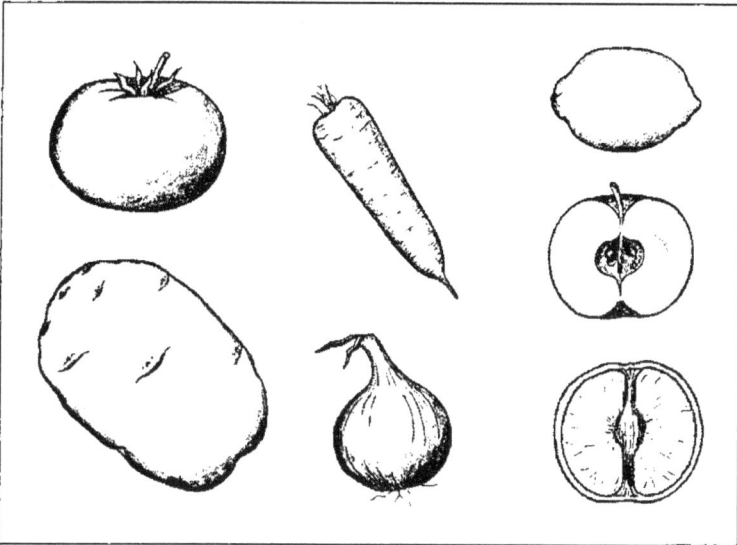

Some fruits and vegetables that contain vitamin C

whether they are eaten raw or cooked. Cooked potatoes have some of this vitamin. Many other vegetables and fruits contain helpful amounts of vitamin C.

We have told you that this substance prevents scurvy.   It is also necessary for growth. You need it, too, in order to have healthy teeth and gums.

Vitamin D

We now come to the famous vitamin that is sometimes called the sunshine vitamin.

If enough sunshine touches your skin, you secure all the vitamin D you need without eating any. Vitamin D is formed in the tissue beneath the skin in the presence of sunlight.   It is the rays of light having short waves that help us in this way.   They are called ultra-violet rays.

As you may have heard, certain electric lamps are now made that give off ultra-violet rays. They are used to help people who do not get enough sunshine. Such lamps are dangerous to play with. They should be used only by people who understand how to handle them. It is harmful to get too much ultra-violet light.

Scientists have learned that vitamin D is formed in certain foods when they are exposed to ultra-violet rays. There is a little vitamin D in milk and butter. There is more in egg yolk and some fish, such as sardines and salmon.

Cod-liver oil has more of it than any of our ordinary foods. If children who do not get enough sunshine are given cod-liver oil, they do not have *rickets* (a disease causing weak and crooked bones). People learned long ago that cod-liver oil is good for children, but they did not know until recently why it is healthful. We know now that it is the vitamin D in this oil that helps prevent rickets.

For a while scientists knew of no substance richer in vitamin D than cod-liver oil. Then they discovered that halibut-liver oil has very much more of both vitamins A and D than

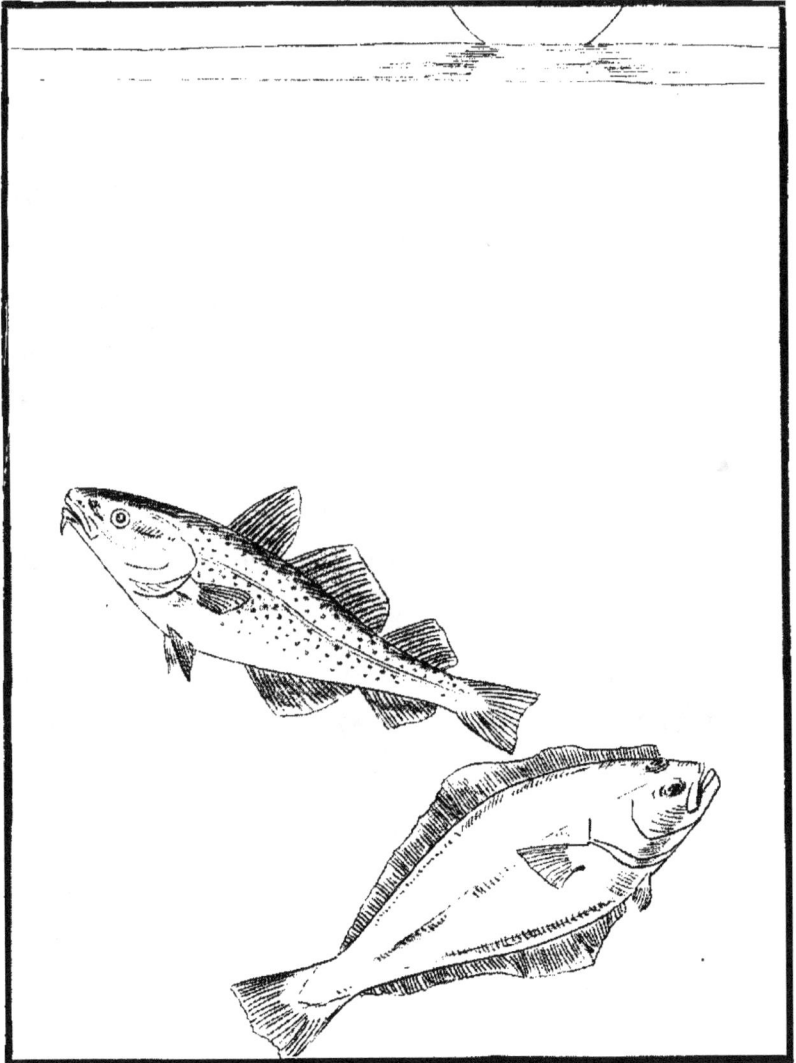

We may get vitamin D from sunshine or by taking
oil of certain fishes.

cod-liver oil has. They made this discovery shortly before this book was written. So we are able to tell you about it while the information is still new.

QUESTIONS AND ACTIVITIES

Look for the word *milk* on the pages of this book that concern food — pages 365–421. Copy on a piece of paper all the sentences that have the word *milk* in them. Does this help you to understand why milk is sometimes called a "perfect food"? Write a short essay about milk. In your essay tell why children are given so much milk to drink.

Did you ever see a citrus tree growing? If you never did, plant a few citrus seeds if you can get them. What fruits would you cut open for citrus seeds? Plant your seeds in a flowerpot or a tin can. Notice what sort of leaf your young plants have.

What vitamins would you get if you should drink some *orange nog?* (This might be made by beating together a cup of milk, half a cup of orange juice, and about two tablespoons of brown sugar.)

You may look at the pages on vitamins to see what vitamins there are in orange juice and milk — if you do not remember.

What is meant by "whole wheat"? Explain why bread made with whole-wheat flour may be a better food than bread made from white flour. What is bran?

What vitamin helps children to have healthy eyes? What vitamin is needed to help make strong bones? Why is "sunshine vitamin" a good name for one of the vitamins? Which vitamin is this?

Read about short light waves in *Through Four Seasons,** Chapter Twelve. Is it possible to get too much ultra-violet light from sunshine? Is there danger of getting too much from "sun lamps"? What is meant by *sunburn?* Can ultra-violet light pass through ordinary window glass? Is it stopped by smoke and clouds?

Tell how each of the vitamins A, B, C, and D helps to keep you well.

An apple tree belongs to the Rose Family. A tomato plant belongs to the Nightshade Family. To what family does an orange tree belong?

* See page 424 for book list.

Explain how an apple, an orange, and a tomato differ in certain ways, such as their skins, pulps, and seeds. Name one vitamin that occurs in all three of these fruits.

Did you ever see a hen or any other feathered animal take a sun bath? What good would a sun bath be to a robin? Did you ever see a dog stretched out in the sunshine? Explain how a cow or other furry animal might get vitamin D.

BOOKS TO READ

Selections from several books have been suggested in connection with different subjects in this book. Here is a list of the books to which such reference has been made:

The Bee People, by Margaret Morley (McClurg)

The Book of Woodcraft, by Ernest Thompson Seton (Doubleday, Doran)

Cotton and Other Useful Fibers, by Nellie B. Allen (Ginn)

First Lessons in Nature Study, by Edith M. Patch (Macmillan)

Forest Facts for Schools, by Charles Lathrop Pack and Tom Gill (Macmillan)

Hexapod Stories, by Edith M. Patch (Little, Brown)

Holiday Hill, Holiday Meadow, and *Holiday Pond*, by Edith M. Patch (Macmillan)

Introduction to World Geography, by Philip A. Knowlton (Macmillan)

Our Plant Friends and Foes, by William Atherton Du Puy (Winston)

Soap Bubbles, by Ellen McGowan (Macmillan)

The Spring of the Year, Summer, The Fall of the Year, and *Winter,* by Dallas Lore Sharp (Houghton Mifflin)

Surprises, by Edith M. Patch and Harrison E. Howe (Macmillan)

Through Four Seasons, by Edith M. Patch and Harrison E. Howe (Macmillan)

Trees, by Julia E. Rogers (Doubleday, Doran)

The Work of Scientists, by Edith M. Patch and Harrison E. Howe (Macmillan)

Working with Electricity, by Katharine Keelor (Macmillan)

INDEX

Acids, 245; and alloys, 241; and copper, 242–243; hydrochloric, 383; vegetable, 240; vinegar, 108, 242, 244

Adobe, 70, 71, 79

Africa, 55, 278

Air, conditioning, 204; hot and cold, 178–208; necessary to fire, 203; pressure, 141, 142, 145, 199, 259, 261, 263–264

Air-lift, or suction, pump, 140–142, 143, 264

Airplane duster, 285, 287

Alabama, 284, 286

Alaska, 362

Albumen, of egg whites, 381

Alcohol, in varnish, 117

Alder, woolly aphid on, 126

Alloy, 135, 240, 241

Alpaca, 354, 355

Aluminum, dishes, 239–240; paint, 104; strength of, 212

Ammonia, 186

Angora goat, 355–356

Animals, dormant, 178–179; electricity in, 171–175; herbivorous, 377

Ant, and honeydew, 372, 373; carpenter, 83

Antarctica, 293

Anthracite, 191

Anti-Steel-Trap League, 361

Aphid, 125, 281, 373; woolly, 125–126

Apple, blossom visited by humming bird, 32; fruit contains vitamin C, 417; wax on fruit, 123; woolly aphid on, 126

Arbor vitæ, 98

Arctic region, 43, 338, 371

Argentina, 300, 301

Arizona, 272, 273

Artesian well, 138, 139

Ash, tree, aphid on, 126

Ashes, in all animals and plants, 391–392; sewage

drained through, 154; wood, used in making soap, 249

Asia, 10, 119, 255, 311

Asia Minor, 356

Asparagus, 395, 409

Aster, 18; seed, 274

Astrakhan fur, 352

Aurora, 175–176

Australia, 278, 358

Aviator flying suit, 339, 340, 341

Ax, 84, 221, 222, 223

Bacterium (and bacteria), 156; causing diseases of cotton and potato, 279, 290; help the retting of flax, 297; in dust, 246; in sewage, 148–149, 154; in spoiled food, 248

Bagworm, Abbot's, 83, 84

Balance, a device for measuring weight, 231, 232, 244; as lever, 232, 234

Balance pole, or well sweep, 139, 140

Balmony, or turtlehead, 16

Balsa, a soft wood, 86

Balsam, Canada, 113

Balsam fir tree, 113

Bamboo, 45–46, 49; silicon in stalk, 48

Banana, 417

Barley starch granule, 374

Bat, dormant, 178

Bayberry, wax on fruit, 123

Beans, contain iron, 395; contain vitamin B, 410; protein food, 381; rich in nitrogen, 390; scarlet runner, attractive to humming birds, 33; starch granule, 374, 376

Bear, cave, 50; dormant, 178

Beaver, befriended by Grey Owl, 347–349

Bed rock, 137, 139

Bee (see also Bumblebee, Carpenter bee, Honeybee), 23–24, 29, 36–37, 83, 372, 373

Bee balm, 34

Beech wood as fuel, 188

Bee glue, 130

Beeswax, 121, 128–129

Beet, sugar, 371; water content, 398

Beetle, injurious to cotton, 281, 283; pollen eater, 30; wood borer, 82
Beriberi, 408
Bering Sea, 362
Berries for birds, 35-36
Birch tree, 97, 100
Birds, berry eaters, 35-36; builders with grass, 45; carpenters, 81-82, 83; masons, 61, 62, 65, 79; oiling feathers, 124-125. *See also* Cliff swallow, Duck, Goldfinch, Grosbeak, Humming bird, Robin, Woodpecker
Bit, a tool with screw action, 221, 224
Bituminous coal, 191-193
Black arm, 272, 279
Blackleg, 279
Blade, leaf, 47
Blueberry, wax on fruit, 122, 125
Board, building, 91-93, 207
Body builders (food), 385-400. *See also* Minerals, Proteins, Water
Boll, cotton, 274, 275
Boll weevil, 283-287

Bollworm, 282
Bolt and nut, 226, 235
Bones, 392-393
Botanist, 299, 300, 301
Boy Scouts, Sixth Law of, 346, 361
Bran, 409
Brass, 120, 135
Brick, 58, 61, 70-73; pressed, 72, 79; scouring, 251; sun-dried, 70
Bridal wreath, 18
Brideweed, 9
Brilliantine, 356
British Isles, 5, 255
Broadleaf trees, 100-101
Broom, a shrub, 253-255
Broom corn, 255-257
Brooms and brushes, for cleaning, 253-257
Buckwheat starch granule, 374
Building board, 91-93, 207
Bulk (volume), devices for measuring, 232, 233
Bumblebee, 37, 38; home of, 45; visits butter and eggs, lilac, snapdragon, turtlehead, 9, 11, 17, 24
Bumblebee flowers, 11, 24

Bumblebee moth, 24

Butter, a fat, 380; contains vitamin A, 405; vitamin D, 419

Butter and eggs, a plant, 7, 9–11

Butterfly, Baltimore, 17; swallowtail, 21–23

Buttermilk, 380

Cabbage, contains the minerals calcium and phosphorus, 395; contains vitamin C, 417

Cable, with electric wires, 167–168

Cactus, chlorophyll in stem of, 368

Caddis worm, as a mason, 63–64, 65, 79

Calcium, a mineral in bones, 392–393; present in vegetable foods, 393, 394, 395

Calfskin for aviation suit, 340

California, 12, 317, 413

Camel (and camel's hair), 353–355

Canada, 347

Canada balsam, 113

Candles, 159, 161

Carbon, chief substance in charcoal, 189; in coal and coke, 190–191; in fat, 379; in protein food, 388; in starch and sugar, 375, 376; with oxygen, furnishes energy, 366, 367, 368, 369, 379, 382

Carbon dioxide, escapes in air from fire and respiring animals and plants, 369–370; formed when fat burns, 379; when proteins burn, 382; removed from city water supply, 153; CO_2, 370, 379, 382

Cardinal flower, 33

Carpenter, 81–101; tools of, 211–228

Carpenter ant, 83

Carpenter bee, 83

Casein, of milk curds, 381

Cast iron, 212, 239

Caterpillar, 319; builds log cabin, 83; injurious to cotton, 282; obtains cellulose, 324, 329; paints inside of cocoon, 126; silk of, 308–318, 321, 322;

silk glands and spinnerets, 318

Catfish, electric, 173, 175

Caustic lye, or caustic potash, 263

Caustic soda, 333, 334

Cave, 50; bear, lion, hyena, 50; man, 51, 68

Cecropia moth, 308

Cedar oil, for cleaning, 258

Celery, 395

Cell (simple part in animal or plant body), 148, 324, 329, 386, 388

Cellulose for silk and rayon fibers, 323–334

Cement, used by mason, 61, 73, 74–75

Central America, 267

Cereals contain vitamin B, 409. *See also* Rice (whole grain), Wheat (whole)

Cesspool, 148

Chalk, 74; dust, 242

Char, charred wood, 189

Charcoal, filter for water supply, 146, 151; fuel, 187, 189–190; sewage drained through, 154

Chardonnet, Count, 326, 327

Chardonnet process of manufacturing rayon, 327

Cheese contains protein, 389; vitamin A, 405

Chemist, 110, 120, 239, 240, 250, 323, 325, 401

China, 94, 278, 305, 310, 311, 314–316, 318

Chips, made by wood-cutting tools, 220–224

Chisel, a mason's tool, 77

Chlorophyll, 368, 369, 370

$C_{12}H_{22}O_{11}$, sugar, 370

Chromium, a metal, 241

Circuit, electric, 160, 161

Cistern, 145–146

Citron, a citrus fruit, 415

Citrus fruits, contain vitamin C, 416; trees of the Rue Family, 412–415. *See also* Citron, Grapefruit, Lemon, Limes, Orange

Clamp, a device that works with screw action, 226

Clapboard, 106, 107

Claw hammer, 212–218

Clay, 137; brick, 70–73; building of mason wasp, 64; cement, 74; dishes, 238; nest of cliff swallow, 61–62, 65, 78, 79; rammed earth, 55, 56; terra cotta, 73

Cleaning devices and materials, 242, 246–264

Clearwing moth, 23–24, 25, 37, 38

Cliff dweller, 52, 53, 68

Cliff swallow as a mason, 61-62, 65, 78, 79

Clock, 233

Clover, 405

CO_2, symbol for carbon dioxide, 370, 379, 382

Coal, 190–194, 366, 367

Coal oil (kerosene), 157

Coal tar colors for dyes and paints, 110

Coconut oil, 250

Cocoon, a rain-proof chamber, 126–127; giant silkworm, 310–312, 319; mulberry silkworm, 312, 313, 314

Codfish, 407

Cod-liver oil, contains vitamin A, 407; vitamin D, 419

Coke, 190, 194

Columbine blossoms, 34

Composite Family, 18

Compound and element, chemical, 48

Compressed-air tank for water, 144, 145

Concrete, 73–76; basin for water reservoir, 151; cistern, 145; tank for sewage, 154

Conduit for electric wire, 164, 167–168

Conifer, 98

Conservationists, 343

Cooks' tools, 229–245

Copper, dishes, 242–243; pipes, 135; and verdigris, 243, 244; sheets on roof near chimneys, 103, 104; wire for electricity, 164

Corkscrew, 235

Corn, popped kernel, 384; starch granule, 374

Corn earworm, 282

Cornel bush, 36

Cornstalk, blade, 47; heat insulator in building

board, 92, 93; node, 47, 59

Cornstarch, 384

Cornus bush, 36

Cotton, 266-290; boll, 274, 275; diseases of cotton plant, 278-281; fibers, 270, 273-276, 330, 337; insects helpful to and injurious to cotton plant, 281-289; linters, 330; mercerized, 333; varieties (Egyptian, Hopi, Sea Island, upland), 270-273

Cotton Belt, 269, 279, 284, 296

Cotton gin, 272, 276-277

Cottonseed oil, food, a fat, 381; used for soap, 250

Cotton trade, 277-278

Cotton trees and shrubs, 267-269

Cranberry, high-bush, 36

Crank, a device with screw action, as a lever, 236-237

Crazing, of dishes, 239

Cream, rich in fat, 380

Crude oil (petroleum), 195

Crystals, of verdigris, 243

Cucumber plants troubled by eelworms, 280

Cup, for measuring bulk, 232, 233

Curd, milk, 381

Current, electric, 160, 164-166, 195; hot-air, 201; water, 201-202

Cynthia moth, 311

Dahlia, 18

Daisy, 18

Davis, Thomas, 20

Day lily, 18, 34

Digitalis (foxglove), 5

Disease germs in water, 136

Diseases, of cotton plant, 278-281; cucumber, 280; flax, 298-299; peas, 280; potato, 279; silkworm, 323. See also Beriberi, Rickets, Scurvy, Typhoid fever

Dishes to stand heat and cold, 238-240

Distilling oil from turpentine, 113-115

Dogwood berries, 36

Dormant animals, 178-179, 206

Driers, in paint, 113, 116

Dropper (pipette), 258–260, 261

Duck, 125

Dugout, 53

Dust, irritating, 246

Dusting, citrus orchard, 415; cotton field, 285, 286, 287

Earth, colored, as pigment for paint, 110; layers of earth surface, 136–138; materials for shelter, 50 - 58, 61–80

Earth oil (petroleum), 195

Earthquake, 137

Eel, electric, 172, 174

Eelworm, 280–281

Egg, white of, 381, 389

Egg, yolk of, contains fat, 380; iron, 395; vitamin A, 404; vitamin D, 419

Egypt, 66, 70, 94, 291, 293

Elderberry, 35

Electric, cable, 167–168; circuit, 160–161; current, 160, 164–166; energy, 165; engine, 143; fuse, 164–166, 167; insulation (and insulator), 163–164, 167; light, 157, 159–161; motor, 144; shock, 161–163; switch, 160–161, 164; wire, 158, 164–170, (grounded, 163, 168–170)

Electric catfish, 173, 175

Electric eel, 172, 174

Electrician, 157–177

Electricity, 159–177; in animals, 171–175; used for heat, 165, 170, 195–196

Electric ray fish, 174, 175

Element and compound, chemical, 48

Energy, electrical, 165; heat, 366–368, 376; motion, 367, 368; used by human body, 165, 366–368, 376

Engine, electric, 143; gasoline, 143

England, 55, 276, 343

Equisetum (horsetail, scouring rush), 251–252

Eskimo, 41, 43, 54, 338, 339, 341

Europe, 5, 8, 10, 12, 13, 41, 189, 255, 268, 311

Evaporation, 179–180, 204–205, 328

Evening primrose, 26, 37

Evergreens, 98–99

Ewe, mother sheep, 357

Fahrenheit, Gabriel Daniel, 184, 186

Fahrenheit thermometer, 183–187

Fats, 378–381. *See also* Butter, Cottonseed oil, Cream, Egg yolk, Lard, Milk, Nut, Olive oil, Peanut butter, Suet, Tallow

Faucet handle, 235, 236

Feathers, oiled, 124

Feldspar, a mineral, 251

Fibers, 303. *See also* Cotton, Flax, Rayon, Silk

Fibroin, sticky fluid in silk gland, 323, 324, 325

Fig tree with lac insects, 119

Figwort Family, 7–8, 16

Filament, in electric bulb, 160

Filter, for sewage, 154; for water supply, 146, 151

Fire-bush, 18

Fireplace, 197

Fire tongs, 234, 235

Fish, a protein food, 381; electric, 172, 173–175; how fish obtain vitamin A, 405–407. *See also* Catfish, Codfish, Eel, Halibut, Ray, Salmon, Sardine, Torpedo

Flax, 112, 292–303; blossoms, 294, 303; fiber, 294, 295–298, 329; retting, 297–298; seed, 112, 293, 294, 295, 299, 300

Flax wilt, 298–299

Fleece, 353–359

Flies as guests of flowers, 30

Flint stone, 48

Florida, 115, 270, 413

Flowers, 2–37

Food, 363–423

Force, resisting and working, 216–220, 227–228, 237

Fountains in reservoir, 151

Fowl, protein food, 381

Fox farms, 350, 351

Foxglove, 2, 5–8

France, 55, 305

Franklin, Benjamin, 197–198, 317

Franklin stove, 197–198

Freezing point, 186

Fruit, *see* Apple, Banana, Bayberry, Berries, Blueberry, Citrus fruits, Cranberry (high-bush), Elderberry, Grape, Raspberry, Strawberry, Tung, Watermelon

Fuel, 187–195

Fuel foods, 364–384; fats, 378–381; proteins, 381–382; sugars and starches, 368–377; used for heat and other body energy, 366–368, 379

Fuel oil, 195

Fulcrum, the support about which a lever turns, 215–220, 227–228, 237

Fungus (and fungi), 246, 279, 290, 370. *See also* Mold, Mushroom

Fur, 338–353, 361, 362; astrakhan, 352; karakul, 352–353; Krimmer, 352; lamb, 340; Persian lamb, 352

Fur farm, 349–353

Furnace, 197, 201–203

Fur-seal treaty, 343

Fuse, electric, 164–166, 167

Galileo, 183

Galls on cotton roots, 279–281

Gardens, flower, 2–38

Gas, fuel, 194–195

Gasoline, 195, 366, 367

Gasoline engine, 143

Gastric juice and starch, 383

Georgia, 114, 270

Geranium, 32

Germ, of seed, 409

Germs, disease, in surface water, 136

Giant silkworms, 308–312, 319. *See also* Cecropia, Cynthia, Luna, Promethea moths

Gimlet, 221, 224

Gin, cotton, 276, 277

Glands, milk, 405; oil, 124–125; silk, 306, 318, 321, 323, 324; swollen glands in neck (goiter), 397; wax, 119, 125

Glass dishes, 239
Goat, Angora, 355–356
Goiter, 397
Goldfinch, 14
Granite, 66–67
Grape, wax on, 122, 125
Grapefruit, 412
Grass Family, 46, 255
Grasses, as building materials, 44–49; contain vitamin A, 405
Gravel and sand, 137
Grease, for soap, 249, 250
Greeks, 14, 41
Greek temple, 66
Grey Owl, 347–349
Grosbeak, evening, 36; pine, 36
Gum, 117, 119, 314, 322
Gypsum, a mineral, 74

Hairs, as fibers in plaster, 73–74; oil present on, 124; surface of, 358, 359
Halibut, 407
Halibut-liver oil, contains vitamin A, 407; vitamin D, 419
Hammer, 77, 212–218
Hannibal, 55

Hawk moths, 25
Hay, 377
Hemp, 303
Hides, for shelter, 40, 41
H_2O, water, 370, 383
Hollyhock, 5, 32
Honey, 371, 373
Honeybee, 24, 129, 130
Honeydew, 373
Honeysuckle, trumpet, 34; twin-flowered, 35
Horsetail (scouring rush), 251 252, 263
Humming bird, 14, 32–34
Humming-bird moth, 25–29, 33
Hydrochloric acid, in gastric juice, 383
Hydrogen, 190; in fats, 388; proteins, 388; sugars and starches, 375, 388; water, 370

Ice walls, 42
Igloo, 43; log, 44; snow, 42; sod, 54
Impudent lawyer, plant, 9
India, 94, 205, 268, 291
Indianapolis, 150
Indian Ocean, 304

Indians, North American, 41, 51–53, 273, 347

Indian's plume, plant, 34

Indigo, pigment, 110

Insects, carpenters, 82–84; helpful to cotton plants, 287–289; injurious to, 281–287; masons, 63, 64, 65. *See also* Ant, Aphid, Bagworm, Bee, Boll weevil, Bollworm, Bumblebee, Bumblebee moth, Butterfly, Caddis worm, Caterpillar, Clearwing moth, Flies, Hawk moths, Lac insect, Leaf-cutter bee, Mason wasp, Mosquitoes, Scale insects, Silkworms

Insulation (and insulator), for electricity, 163–164; for heat, 93, 207–208

Iodine, 396–397

Ireland, 193, 291, 293

Irish linen, 291

Iron, alloys, 240, 241; in food, 395; in water supply, 153; pipes of, 135; protected by lacquer, 120; protected by paint, 104; rust, 102, 103, 120, 153, 240, 241

Iron, cast, 48, 212, 239

Iron oxide, or rust, 102, 241

Italy, 318

Jackscrew, 227

James I, of England, 316

Japan, 94, 120, 318, 343

Japanese quince, 18

Jute, 303

Karakul sheep, 352–353

Kerf, 222, 224

Kerosene, 195; lamp, 157–159

Kid, Angora, 356

Kidneys, action of, 398

Kiln, for brick, 71, 72; for cement, 75

Kindness to animals, 346–349, 359, 361

Kitchen tools, 229–245

Knife, strokes of, 221, 223

Krimmer, fur, 352

Lac, a wax, 119–120

Lac insect, 119

Lacquer, 119, 120–121

Lacquer tree, 120

Lambs, and lamb fur, 340, 352, 357

Lamp, kerosene, 157–159

Lanolin, wool wax, 124

Larch, 98

Lard, 380

Larkspur, 5

Lead, used by plumber, 133, 135

Lead, red, a paint, 104

Lead, white, a paint, 108, 110, 116

Leaf-cutter bee, 30 32

Leaves, contain vitamin A, 404–405. *See also* Clover, Grass, Lettuce, Spinach

Legume, 390

Lemon, citrus fruit, contains vitamin C, 410, 411; juice prevents scurvy, 415; waterproof rind, 124

Lentil, rich in nitrogen, 390

Lettuce, contains calcium and phosphorus, 395; vitamin A, 404

Lever law, 216, 218, 219, 220

Levers, of the first class, 220, 234; of the second class, 234; of the third class, 235. *See also* Balance, Claw hammer, Crank, Jackscrew, Nutcracker, Potato ricer, Scissors, Seesaw, Shears, Snips, Tongs, Well sweep

Lichens, on unpainted wood, 105

Lightning, 175–176

Lights, 157–161

Lignin, 332

Lignite, 193

Lilacs, 21–24

Lily Family, 18

Lily of the valley, 18

Lime, 399; contains calcium, 392 393; used in acid soil, 393; in plaster, 73; in sewage, 154

Limes, citrus fruit, contain vitamin C, 416; used to prevent scurvy, 416

Limestone, a building stone, 67; and marble, 67–68; contains calcium, 392–393; used in cement, 74

Linen, 291 303, 329

Linnet, 302

Linoleum, 302

Linseed, flax, 112, 302

Linseed oil, 104, 111–112, 116, 117, 295

Lint, cotton, 270

Linters, cotton, 270, 329, 330

Linum, flax, 112, 302

Livers, contain vitamin A, beef, 405; chicken, 408; fish, 407

Llama, 355

Log cabin, 83, 84–86

Log igloo, 44

Long-leaf pine, 115

Louisiana, 146, 284

Lumber, from broadleaf trees, 100–101; from conifers, 98–100; preparation and use of, 86–91

Luna moth, 308

Lye, caustic, 263

Madagascar, 304, 305

Maine, 12, 109

Mammals, body temperature of, 178–180

Maple, 188; lumber, 100; sap, 371

Marble, 67–68

Marigold, 18

Maryland, 132

Mason wasp, 64–65, 79

Mason's tools, 76–79

Mason's work, 50, 61–80, 93–94

Mass, 213–214

Measuring devices, to measure bulk (volume), length, temperature, time, and weight, 231–233. *See also* Balance, Clock, Cup, Rule, Scale, Spoon, Thermometer, Watch

Meat, body builder, 389; contains iron, 395; fat, 380; fuel food, 381

Meat chopper, 236

Mercer, John, 333

Mercerized cotton, 333

Mercury, 184, 186

Metal, conductor of heat, 93; conduits or tubes, 167; discolored, 240–243; path for electricity, 164; pipes, 135; powders for cleaning, 251; protected by lacquer, 120; tank, 145. *See also* Aluminum, Brass, Chromium, Copper, Iron, Lead, Mercury, Nickel, Silver, Tin, Zinc

Mexico, 66, 67, 283

Milk, 421; contains calcium, 393; fat, 380, 381; iron, 395; protein, 381; vitamin A, 405; vitamin D, 419

Milk curd, 381

Milk sugar, 373, 381

Mill, pulp, 331–332; sawmill, 87–88

Mineral fuel, coal, 190

Mineral oil, petroleum, 195

Mineral salts, 392

Minerals in food, 386, 391–397. *See also* Ashes, Calcium, Iodine, Iron, Mineral salts, Phosphorus

Mississippi, 284

Mohair, 355, 356

Mold (fungus), 246, 248, 262–263, 377

Mole, living in dugout, 53

Momentum, 213–214

Mortar, 73–79

Mosquitoes in cisterns, 146

Moss on unpainted wood, 105, 107

Moth, or larva of moth, 308, 309. *See also* Bagworm, Bollworm, Giant silkworms, Hawk moths, Silkworms

Motor, electric, 144

Mouse, field, 45

Mucous membrane, 247

Mulberry (leaves and tree), 312, 316, 317, 318, 329

Mullein, 7–8, 12–15

Mushroom (fungus), 370

National parks, 95

Nectar, 9, 11, 23, 29–30, 32, 34

Nephila spider, 304–306

New England, 68

Nickel, 241

Nitrogen, 190, 388–390, 399

Node of stem, 47, 59

Nonconductor, 92–93, 207–208. *See also* Insulation

North Dakota, 302

Northern light, 176

Nut, fat food, 381; protein, 381, 390

Nut, metal device, 226, 235

Nutcracker, as lever, 234

Oak, for lumber, 97, 100; white, 100, 101

Oat, starch granule, 374

Oatmeal, iron in, 395

Oil, 107–108; for fuel, 195, 196; in sap of pine trees, 122; in soap, 250; on feathers and hair, 124. *See also* Cedar, Cod-liver, Cottonseed, Halibut-liver, Olive, Poppy-seed, Tung

Oil lamp, 157–159

Ointment, 124

Olive oil, a fat, 381

Onion, contains vitamin C, 417; starch granule, 375

Orange, citrus fruit, climate for, 413; contains vitamin C, 410; juice to prevent scurvy, 415; water content, 398; waterproof rind, 124

Orange nog, 421

Oregon, 296, 361

Oswego tea, 34

Oxygen, combines with carbon to form carbon dioxide, 369, 382; combines with hydrogen to form water, 370; combines with iron, 102, 153; gas given off when coal was forming, 190; hardens paint, 112, 113; present in air, 182, 205, 367, 389; in proteins, 388; in starch and sugar, 375

Oyster-shell scale, 414

Paint, 102–116. *See also* Aluminum, Red lead, White lead

Painter, work of, 102–130

Paraffin, 121, 123

Pasteur, Louis, 323, 326

Pea Family, 253, 381, 390

Pea, starch granule, 374, 376

Peanut (and peanut butter), protein food, 382, 390; starch granule, 376

Peas, 280; contain iron, 395; protein, 381, 390; vitamin B, 409; disease of, 280

Peat, a fuel, 193

Pennsylvania, 317

Persian lamb fur, 352

Perspiration, 398

Peru, 354

Petroleum, 195

Petunia, 25–29

Philadelphia, 10

Philippine Islands, 49

Phosphorus, 399; a mineral in bones, 393

Pigment, coloring matter in paint, 110, 111, 116

Pilgrims, 68

Pine, yields timber and turpentine, 97, 113, 114, 122. *See also* Long-leaf, Short-leaf

Pine Family, 98–100

Pipes, materials for, 135; uses for, 115, 133–134, 140–142, 143, 144, 149, 150, 153

Pipette, or dropper, 258, 260, 261

Pistil of flower, 29, 288

Pitch, burns readily, 187; covers wound in pine tree, 122; in knot, 119

Pith, in stalks, 46

Plane, a carpenter's tool, 94

Plant materials for walls, 43–49

Plants, 3–37; with useful fibers, 303. *See also* Cereals, Composite Family, Cotton, Figwort Family, Flax, Fruit, Fungus, Grass Family, Horsetail, Leaves, Lichens, Lily Family, Moss, Pea (or Pulse) Family, Rose Family, Sumac, Trees, Vegetables

Plaster, 73–74, 76

Plaster board, 76

Plum, fruit with wax, 122

Plumber, work of, 133–156

Plush, 356

Pollen, 11, 29–30; bearers, 30, 32, 288–289

Pongee silk, 312

Poppy-seed oil, 117

Potash, caustic, 263

Potassium, 249, 250, 263

Potato, contains vitamin C, 417; diseases of, 280, 289–290; water content, 398

Potato ricer, as lever, 234

Potato starch, 383–384; granule, 374, 375

Pribilof Islands, 344, 362

Priming coat of paint, 111

Primrose, evening, 26, 37

Promethea moth, 308, 309

Propolis, or bee glue, 130

Proteins 381–382, 386–391. *See also* Albumen, Beans, Casein, Cheese, Egg white, Fish, Fowl, Legume, Lentil, Meat, Milk, Milk curd, Nut, Peanut (and peanut butter), Peas, Shellfish

Pulse Family, 376, 381, 390

Pump, connected with compressed-air tank, 144; suction, or air-lift, 140–142, 143, 264

Pyramids, 66

Quince, Japanese, 18

Rabbits, 350, 351

Radiator, 202

Rajah silk, 312

Rammed earth, directions for making, 59-60; walls of, 55-58

Ranstead, 10

Raspberry, flowering, 18; water content, 397

Ray, electric fish, 174, 175

Rayon, 163, 321–337

Réaumur, René de, 321

Red-lead paint, 104

Register, opening of hot-air pipe, 201

Reptile, dormant, 179

Reservoir, for water, 151

Resin (rosin), 113–115, 117, 122, 130, 187, 322, 332

Retort (oven), 189, 195

Rhode Island, 152

Rice, polished and whole grain, 408; starch granule, 374, 375; vitamin B, 408

Rickets, 419

Robin, picking berries, 35

Rocks, 58; crushed for cement, 75; pebbles used by caddis worm, 63; silicon in, 48; walls of, 50

Rock wool, 208

Romans, 14, 41

Rome, ruins in, 66

Roof, lead, 133; shingled, 54, 105; thatched, 45; tile, 56–57

Root knot, 279–281

Rose, 18, 31

Rose Family, 18

Rosette of mullein plant, 12, 13, 14

Rosin (resin), 113

Rubber, for insulating electric wires, 163

Rubber tree, 127

Rue Family, 412

Rule, ruler, measuring, 233

Russia, 343, 358

Rust, iron oxide, 102; in water pipes, 153; on kitchen utensils, 240, 241; prevented by lacquer, 120; prevented by paint, 103-104

Saliva changes starch, 376

Salmon, 419

Salt, used in soap, 249

Sand, composed largely of silicon, 48; in cement, 75; in forming rocks, 58; in layer of earth surface, 137; in plaster, 73; in rammed earth, 55-56; silica ware made from, 239; used for cleaning kitchen utensils, 251; used for water filter, 146, 151

Sandpaper, 94

Sandstone, 67

Sap, of pine, 122

Sardine, 419

Saw, 77, 222-224

Sawdust, chips, 224; insulating material, 208

Sawmill, 87-88

Scale, device for measuring weight, 231, 232, 245

Scale insects, 119, 414

Scale, oyster-shell, 414

Scarlet runner bean, 33

Scissors, as levers, 220

Scouring materials, 250-253

Scouring rush, 252

Screws, devices with screw action, 224-227, 235-236; bit, 221, 224; bolt and nut, 226, 235; clamp, 226; corkscrew, 235; crank, 236 237; faucet handle, 235, 236; gimlet, 221, 224; jackscrew, 227; meat chopper, 236; vise, 226; wood, 224

Scurvy, 414-416

Sea animals and plants obtain vitamin A, 405

Seesaw, as lever, 215-216

Sericin, waxlike material in silk fiber, 324, 325

Seton, Ernest Thompson, 359, 361

Sewage, 133, 147–149, 153–154

Shale, used in cement, 74

Shantung silk, 312

Sharp, Dallas Lore, 361

Shears, as levers, 218–220

Sheath, of leaf, 47

Sheep, follow a leader, 342; grazing, 360, 361; hair covered with fatty oil, 124; valued for food and fur, 356–357

Sheep, karakul, 352–353

Sheepskin suits, 340

Shellac, 119–120

Shellfish, protein food, 381

Shell flower, 16

Shelters, 40–58; of earth materials, 50–58, 61–80; of hides, 41; of plant materials, 43–49; of snow and ice, 42–43

Shingles, 105, 106, 107

Shortleaf pine, for timber, 99; for turpentine, 115

Silica ware, 239

Silicon, 263; in sand and certain rocks, 48; in stalks of plants, 48, 252

Si-Ling-Chi, 314

Silk, 304–320, 321; artificial, 327, 335; wild, 310–312. See also Pongee, Rajah, Shantung, Tussah

Silk glands, of caterpillars, 318; of spiders, 306–308

Silkworm disease, 323

Silkworms, Chinese, or mulberry, 312–318, 321, 329; giant, 308–312, 319

Silver, conductor of heat, 92–93; discolored, 241–242

Sisal plant, 303

Skins for shelter, 40, 41

Slag, blast-furnace, 74

Slate, waste, 74

Slaughter yard, laws regulating proceedings, 345

Smoke, 190, 192–193

Snake, dormant, 178

Snakehead, 16

Snapdragon, 7, 8–9, 16

Snips, as levers, 218–220

Snow shelters, 42–43

Soap, 248–250, 251, 263

Sod house, 43, 53–54

Soda, caustic, used in making mercerized cotton and rayon, 333, 334
Sodium, in salt, 249; used in soap making, 249, 250
Soil layer of earth, 136–137
Solder, 135
Soot, 190, 193
Sorghum, 255, 371
South America, 267, 299, 355
South Carolina, 270
South Dakota, 294
Southern light, 176
Spain, 55
Spider silk, 304–308, 319
Spinach, contains calcium and phosphorus, 395; vitamin A, 404
Spinneret, of caterpillar, 318, 321, 322, 325; of spider, 306, 307; machine, 322, 326, 328, 334
Spoon, for measuring bulk, 232, 233
Spring, water, 136
Spruce wood for rayon, 331–335
Spurge Family, 118, 127
Squash, contains calcium and phosphorus, 395

Stamen, of flower, 29
Standpipe for water, 151
Starch, 368–377; granules, 374, 375–376
Steam, 134
Steel, 212, 239; knives, 250–251; stainless, 240–241
Stone and stone buildings, 61, 65–70
Stoves, 197–203
Straw, 45, 59, 92, 93
Strawberry, rich in vitamin C, 417
Suction cleaner, 247, 258–262
Suction, or air-lift, pump, 140–142, 143, 264
Suet, 380
Sugar, 368–377
Sugar cane, 371; stalks used for building, 92
Sugar factory, 368, 371
Sugar tongs, as levers, 234, 235
Sulfur, causing silver to tarnish, 242; dust to kill insects, 414, 415
Sumac, 120
Sunburn, 422

Sunflower, 18
Sunshine vitamin, 418, 419
Surface water, 135–136, 138
Swallow, cliff, as mason, 61–62, 78, 79
Sweat, 398
Switch, electric, 160–161
Switzerland, 327

Tallow, 380
Tamarack, 98
Tank, compressed-air, 144, 145; water, 143–145, 151
Tape, for measuring length, 233
Tarnish on silver, 242
Temperature, 178–208, 233, 341; of body, 178–181
Temples, Greek, 66; in Mexico, 66, 67
Terra cotta, 73
Texas, 115, 283, 284, 355, 356
Thermometer, 180–187, 206, 233
Thinners, for paint, 113–115
Thread, of screws, 226
Thrips, 414
Tiger lily, 18

Time, devices to measure, 233
Tin, in solder, 135
Tinners' snips, as levers, 218–220
Toad, dormant, 178
Toadflax, 9
Tomato, contains vitamin B, 409; vitamin C, 417; water content, 398
Tomato worm, 282
Tongs, as levers, 234, 235
Tools, of carpenter, 210–228; of cook, 229–245; of mason, 76–78; of plumber, 155, 156; wood-cutting, 220–224
Torpedo, fish, 175
Traps, and trapping, 342–345, 346, 347, 350, 359, 361
Trees, 95–101; ancient, 190; broadleaf, 100–101; conifers, 98–100. *See also* Apple, Ash, Balsam fir, Birch, Cotton, Long-leaf pine, Maple, Mul-berry, Rubber, Rue Family, Shortleaf pine, Spruce, Tung

Trumpet creeper, 34

Trumpet honeysuckle, 34

Tung, fruit, 118; oil, 117–119; tree, 117–118

Turk's-cap lily, 18

Turpentine, 113–115, 116, 117, 122; orchard 114

Turtle, dormant, 178

Turtlehead, plant, 8, 16

Tussah silk, 312

Typhoid fever, 136

Ultra-violet light, 418–419, 422

United States, northern, 42, 100, 294, 295; southeastern, 115, 269; southern, 100, 118, 269, 277; southwestern, 51–53, 70; western, 53, 123, 294, 295, 357

Vacuum, 142, 259, 260, 261

Vacuum cleaner, 246, 258–262

Valve, 142, 143

Varnish, 116–121

Vegetables, to prevent scurvy, 413. *See also* Asparagus, Beans, Beet, Cabbage, Celery, Corn, Cucumber, Lentil, Lettuce, Onion, Peas, Potato, Spinach, Squash, Tomato

Velocity, 213-214

Velvet plant, mullein, 13

Ventilation, 203

Verdigris, 243, 244

Vinegar, 108, 242, 244

Virginia, 115, 316

Vise, 226

Vitamins, 401–423; A, 404–408; B, 408–410; C, 410–418; D, 418–423. *See also* Apple, Asparagus, Banana, Beans, Bran, Butter, Cabbage, Cereals, Cheese, Citron, Clover, Cod-liver oil, Egg, Fish, Grapefruit, Grasses, Halibut-liver oil, Leaves, Lemon, Lettuce, Lime, Livers, Milk, Onion, Orange, Peas, Potato, Rice, Salmon, Sardine, Sea animals and plants, Spinach, Strawberry, Sunshine, Tomato, Ultra-violet light, Wheat

Volume (bulk), devices for measuring, 232, 233

Wales, 6, 7, 68, 69
Walls, 41–60
Wasp, mason, 64–65, 79
Watch, device for measuring time, 233
Water, as food, 386, 397–399; supply for household use, 135–155
Watermelon, water content, 398
Water, surface, 135–136
Waterproof coats of animals and plants, 122–127
Water tower, 150
Wax, 119–122; coating on fruit, 122–123; secreted by aphids, 125–126. *See also* Beeswax, Lac, Paraffin
Weevil, cotton boll, 283–287
Weight, device for measuring, 231
Well, artesian, 138, 139; gas, 195; oil, 195, 196; water, 136, 139, 140, 141, 142, 143

Well sweep, 139; as lever, 140, 215, 220, 227
Wheat, starch granule, 374; whole grain contains iron, 395; vitamin B, 409
White-lead paint, 108, 110, 116
Whiting, chalk dust, 242
Whitney, Eli, 276
Wigwam, 40, 41
Wilt, cotton, 279; flax, 298–299
Wood, and the carpenter, 81–101; fuel, 187–190, 366; paints for, 107–109; pulp, 329–332; unpainted, 104–107
Woodchuck, dormant, 178
Woodpecker, 81–82
Wool, 338, 353–362; fiber enlarged, 358, 359
Wool, rock, 208
Wool wax, lanolin, 124
Woolly aphid, 125–126

Zero, 186
Zinc, 103–104, 110